Preserving Food

Pamela Dotter

Macdonald Guidelines

First published 1977
Macdonald Educational Ltd,
Holywell House, Worship Street,
London EC2A 2EN

ISBN 0 356 06019 5

Made and printed by
Waterlow (Dunstable) Limited

Contents

Why preserve?

To ensure a sufficient supply of food throughout the year, man has had to exercise ingenuity to the utmost. Our ancestors had a hard job, but advances in food technology over the last hundred years have made the task of preserving food much easier. We now know why food goes bad; and preventive steps devised in laboratories are easily carried out at home.

Food is subject to nature's laws of sowing, growing and breaking down. To try to prolong the growing season, farmers and gardeners use all their resources of skill in creating a succession of crops and providing fresh produce during the winter. But nature is unpredictable, and we regularly face both gluts and shortages. Marketing specialists efficiently provide luxury, out-of-season fresh foods from other countries and even from the other hemisphere. A craving for strawberries in winter can be satisfied—at a price. Much more gratifying, however, is the chance to have them available prudently preserved.

Self-sufficiency

It is a source of great satisfaction to the frugal housewife to go some way towards being self-sufficient. To those on reduced incomes and pensions it may also be a necessity. The tastes and nutritive value of food preserved at home far outclass those of bought varieties. The fresh flavour of fruit can be enjoyed in home-made syrups, cordials or liqueurs. Instead of spending money on cans of fruit, a jar of bottled fruit has much to commend it on quality as well as price. It is easier to cater for individual tastes and needs. Low-sugar jams suitable for diabetics can be made at home at a fraction of the price asked in the shops. Spare fruits and vegetables can be conveniently used up in chutney. Pickles provide the sweet, sharp and spicy flavours needed to enliven stodgy winter meals.

Presentable preserves

Many preserves make very acceptable, low-cost presents. If Christmas presents are planned ahead they can include home-made luxury jams, jellies and curds. Liqueurs and spirits can be added as in the most expensive bought preserves, to pep up the flavour, and crystallized and glacé fruits can be packed in ornamental boxes.

In the kitchen

Creative cooks can let their imagination run riot when faced with a glut of food. The possibilities of a laden apple tree provide one set of examples. The perfect apples can be sorted out, wrapped and stored in boxes; the remaining large apples should have any diseased parts cut out, and can then be cut into rings and dried; windfalls can be chopped up to make chutney, or purée, or used for bottling or freezing. The small apples can be used to make apple jam, or apple jelly (either on its own, or mixed with mint, or mixed with scarcer, more expensive fruit). The pulp from the jelly can be saved, mixed with spice, and made into apple butter or cheese. If the apples are not keepers, they can be frozen.

The housewife who protests that there is little point in spending hours cooking because the food is all eaten in a flash can take encouragement from the fact that preserves reward her work for a long time.

Preservation through the ages

The original methods of preservation were provided by the natural agents of sun, wind, smoke and salt. Ancient man hung his meat and fish inside his chimney to keep away the flies; and while there it became smoked from the peat fire, which also dried it. Fruits and vegetables were spread out in the sun to dry in much the same way as, throughout hotter regions of the world, they still are in primitive societies.

Salting, smoking and pickling

In former centuries salt, evaporated from the sea, was either rubbed into meat or fish to keep it or made into a brine in which the food was immersed. The food was often smoked as well, to make sure that it could keep for many months without putrefying. This made it hard, and it bore no relation to the delicate smoked meats of today. To make it edible, hours of soaking were necessary. So much salt was used in medieval homes for brining and sousing or pickling that its cost was always listed in the household accounts.

Distributing fish was always a problem formerly, because it was so perishable. Only people living near the sea ever ate fresh fish. Otherwise it was salted, to keep it throughout its long journey inland. The particularly good salt that was available in Portugal was responsible for the fame of Portuguese sardines, though they are nowadays canned in oil. With the salt, they were packed in jars sealed with butter. The Scottish haddock-curers likewise had a

▶ Smoking bloaters, 1882. This process is now mainly used only for flavouring.

▲ The champion bottling team of the United States for 1925.

plentiful supply of good sea salt which helped them establish the quality and acceptability of their product. Arbroath smokies were developed by salting and smoking the fish. The production of these foods is still thriving; but it does so now because of their popular flavours rather than their "keepability". Smoked salmon and trout are now popular luxury foods. Herrings and mackerel are enjoyed in their smoked form for everyday meals. Specialist kitchen shops now sell home smoking kits, with instructions for brining and smoking your own meat, poultry and fish.

In Victorian days, meat was preserved by a combination of salt and saltpetre. Spiced beef was a favourite Christmas joint in homes of that era, when large joints of 10 lb and over were left in a dry pickle containing salt, saltpetre, brown sugar and spices. The pickle added flavour, as well as being to some extent a preservative.

Modern commercial salting is carefully controlled, to give the very mild salt flavour considered acceptable today. Salt now plays very little part in the preservation of bacon, as many more sophisticated methods now exist, from refrigeration to vacuum packaging. In the home, some farmers' wives still cure hams in brine, although it is a lengthy process; and salt is also used domestically in the preservation of runner beans and white cabbage.

Sweet preserves

Sugar was not widely available in Europe until after the eighteenth century; and up to that time, the only sweetening agent in common use was honey. With the availability of sugar, many new preserves were developed. Quinces were made into a sweet preserve which may have been the forerunner of marmalade, since the Portuguese name for quince was *marmelo* and quince preserves were called "marmelade" in cookery books of the day. Jams were so named because the preserves were made by crushing and jamming the fruit with sugar. Other preserves took their names from the consistency of the finished item—jelly, curd, butter and cheese.

Canned foods

The first cans of food were processed at the beginning of the nineteenth century. In the course of their continuous development, patented vacuum cans were produced. Today's modern, automated canning factories use lightweight cans with special linings for fruit and meat. The latest development is aseptic canning, in which high temperatures are unnecessary and the flavours are retained.

Boiling and sterilization

The great breakthrough in preservation came with the discovery by Louis Pasteur that spoilage could be arrested by heat. There are statutory laws for pasteurization of milk, to destroy bacteria carrying disease and to reduce the milk-souring organisms. This process is now applied to many commercial foods. Even wine is subjected to heat treatment, to kill any yeasts that could cause fermentation in the bottle. Sterilized milk keeps indefinitely without refrigeration. The latest development is ultra-heat-treated milk and fruit juices. They are packed aseptically into sterilized cartons which survive months of unrefrigerated storage.

Refrigeration and freezing

In cold climates snow and ice have always been convenient preservers, but formerly there were no means of keeping them. Today, most modern homes have refrigerators, and freezers have become accepted kitchen equipment. It is only when world fuel shortages remind us of the expense of maintaining the necessary low temperature that the half-forgotten methods of preserving—by heat, as in bottling, with sugar, as in syrups, jams and jellies, and with vinegar, as in pickles and chutneys—are revived. Producing them gives the cook the satisfaction of mastering a creative culinary art and avoiding waste when faced with a glut of fresh produce.

▼ An advertisement for an ice-making machine of the late nineteenth century.

Decay versus preservation

Our food is made from organic materials, hence it continually goes through the cycle of development and decay. An apple on the tree first appears green and hard; when ripe it changes colour; and subsequently it gradually shrinks, its skin wrinkles, and it goes dull in colour, eventually showing brown patches. Other changes occur which cannot be seen. All this is caused by enzymes and micro-organisms in and around the fruit. A cut apple, exposed to air, will go brown, and if the fruit becomes very ripe, the enzymes change the pectin to pectic acid. This process is particularly important in jam- and jelly-making.

Micro-organisms

These include bacteria, moulds and yeasts. Some, like those producing penicillin and vinegar, are beneficial; many cause decay.

Bacteria

Some bacteria produce gas and are quite harmless; but others can inflict food poisoning. The harmful ones are Salmonella, Shigella, Staphylococcus, and Clostridium-botulinum. The first two cause typhoid-type diseases, and the rest cause food poisoning. These deadly culprits are invisible, and multiply rapidly in warm temperatures. For this reason food, especially meat, should be heated or cooled very quickly to avoid being kept for long at between 10° and 38°C and 50° to 100°F.

Moulds

Their woolly appearance makes these easily recognizable. They are not harmful, but they do spoil the flavour and appearance of food.

Causes of food spoilage and decay

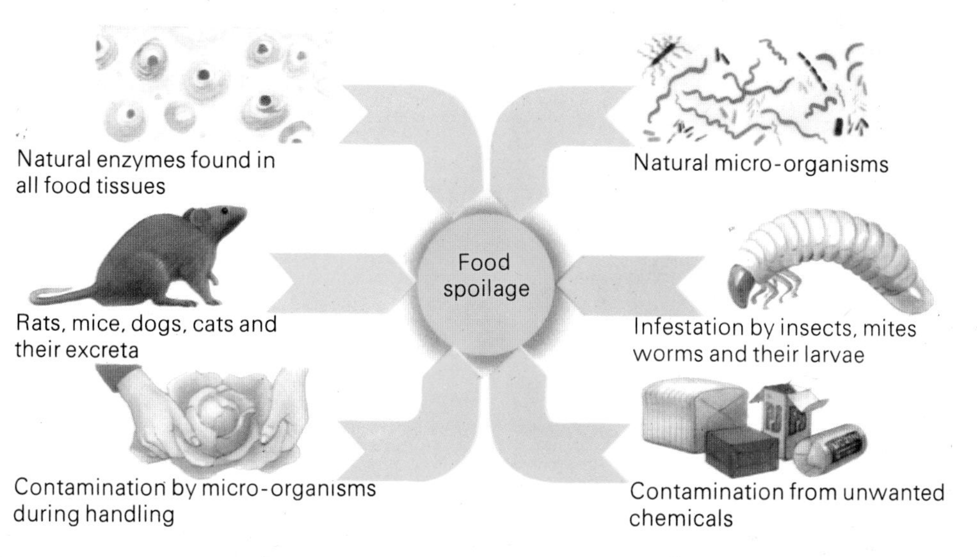

Yeasts

Generally, yeasts are known more for their ability to ferment doughs and beer. They occur on the skins of fruits and are liable to ferment the sugar of the fruit and cause spoilage in bottled fruits and jam.

Heat and cold

Bacteria, enzymes, yeasts and moulds are all destroyed by heat. Fruit, just by being stewed, can be made to keep at least a few days more. For longer storage, food must be processed at higher temperatures, and

Temperature control of food micro-organisms and enzymes

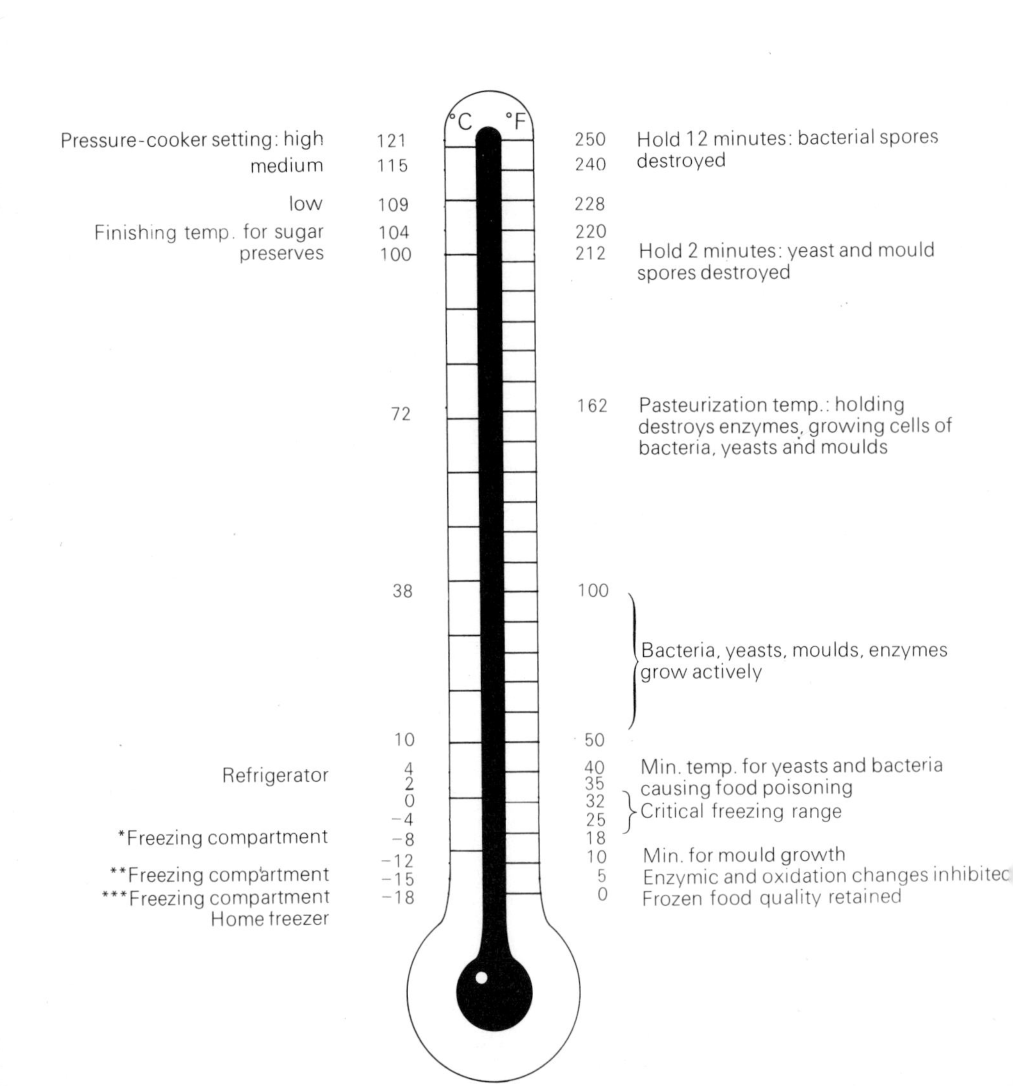

sealed, as in bottling and canning, to avoid contamination from the air. Time must be allowed for heat to penetrate the food.

Cold arrests development of enzymes, moulds, bacteria and yeasts. At around 4°C, 40°F, food keeps for a few days. Lower temperatures, about —8°C, 18°F, make micro-organisms dormant. Temperatures must be as low as —18°C, 0°F, for enzymes to cease food spoilage. Using heat, blanching vegetables for freezing arrests enzyme action. For prolonged storage, food must then be frozen quickly. Blanching vegetables also reduces loss of Vitamin C.

Activity of micro-organisms and enzymes at low temperatures

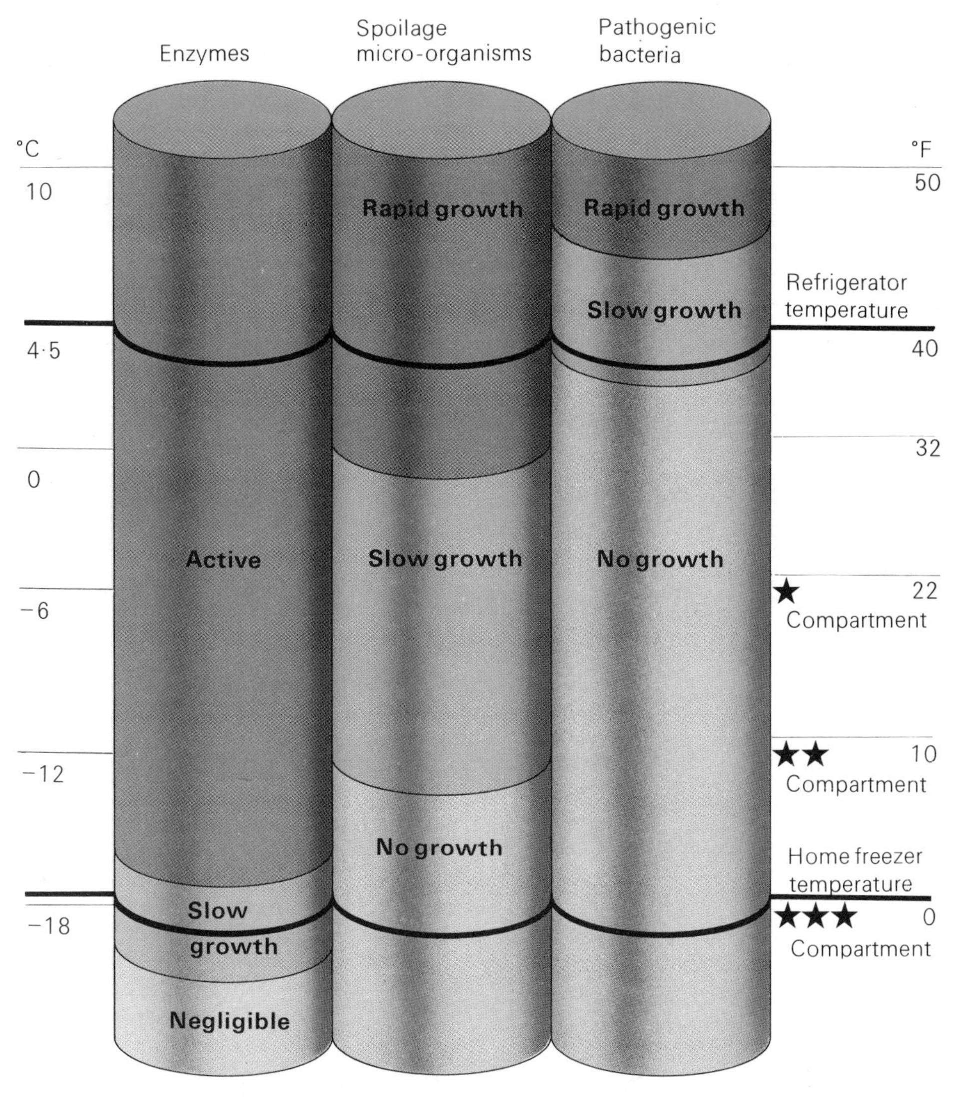

Preserving agents

Sugar, salt, vinegar, alcohol and sulphur dioxide all inhibit the growth of enzymes, bacteria, yeasts and moulds.

Among sweeteners, preserving crystals are useful for jam-making; but granulated sugar can be used instead. In marmalade, brown sugar or black treacle helps to flavour rather than preserve. Golden syrup, glucose, fructose and honey add sweetness; saccharine, however, goes bitter if used for bottling.

In salting food to keep it, rock salt or cooking salt are the ones to use.

For pickles and chutneys high-strength, 8-per-cent, spiced vinegar is the most appropriate; otherwise ordinary table vinegar of 5-per-cent strength can be used.

In preserves using alcohol, brandy goes well with fruit. Vodka, gin, whisky, rum and *eau-de-vie* are often used in fruit liqueurs.

Sulphur dioxide, used for bottling, is available as Campden tablets or sodium metabisulphite. Vegetables cannot be preserved this way, and fruits must be boiled before eating to get rid of the sulphur flavour.

Waterglass, properly called sodium silicate, is dissolved in boiled water to help preserve eggs.

Drying and smoking

Bacteria, yeasts and moulds thrive only if moisture is present. The heat of the sun or an oven is used to remove it from food, which is then protected to prevent its reabsorption from the air. Commercially, food is air – and also freeze-dried.

In smoking food it is the allied salting and drying process that furthers preservation.

▼ Heat, sugar and vinegar are the agents used for these preserves. *Back row:* apple jelly, bottled gooseberries, orange cordial, plum relish, marmalade and raspberry jam. *Foreground:* orange cheese, brown spicy chutney and apricot curd.

The preservation process

Choices of method for preserving vegetables at home

Vegetable	Method	Special points
asparagus	Freeze; bottle	Cut to size of container
beans: runner broad	Freeze; bottle; salt; dry; pickle Freeze; dry	
beetroot	Freeze; bottle; pickle; dry-store; make into chutney	Freeze and bottle very young beetroot
broccoli	Freeze	
Brussels sprouts	Freeze	
cabbage	Freeze; salt; pickle	Use white cabbage for sauerkraut, red cabbage for pickle
cauliflower	Freeze; pickle; make into chutney	Useful for mixed pickles
celery	Freeze; bottle	
courgettes	Freeze, bottle; pickle; make into chutney	Bottle with other vegetables to make ratatouille
gherkins	Pickle; make into chutney	
marrow	Freeze; pickle; dry store; make into chutney	Usually mixed with other vegetables
mushrooms	Freeze; bottle; dry; pickle	
onions	Freeze; dry; pickle; dry store; make into chutney	Use small onions and shallots for pickling
peas	Freeze; bottle; dry	Dry peas in their shells on the plant
peppers	Freeze; pickle; make into chutney	
potatoes: new old	Freeze; bottle Dry store	
sweetcorn	Freeze; bottle; pickle; make into chutney or relish	

Choices of method for preserving fruits at home

Fruit	Method	Special points
apple	Dry; freeze (both as purée and slices); bottle; dry store; use to make fruit butter, fruit cheese, chutney, jam, jélly or juice	Use crab apples for jelly
apricot	Dry; freeze; bottle; use to make jam, fruit butter or cheese, curd, chutney, glacé fruit or brandied fruit	Use dried apricots for chutney and jam
blackberry	Freeze; use to make jam, jelly, fruit butter, syrup, relish or chutney	Use wild blackberries for the stronger-flavoured preserves
black currant	Freeze; bottle; use to make jam, jelly, syrup or liqueur	Process quickly to preserve the Vitamin C
bilberry	Freeze; bottle; use to make jam or jelly	Lacks pectin; mix with apples for jam and jelly
cherry	Freeze; bottle; crystallize; use to make jam or liqueur	Sour varieties (morello) are best. Add pectin for jam
damson	Freeze; pickle; bottle; use to make jam, fruit butter or cheese, relish, chutney or liqueur	
elderberry	Use to make jam or jelly	Mix with apple, blackberry or gooseberry
gooseberry	Bottle; freeze; use to make jam, jelly or chutney	High in pectin; mix with other fruits
grapefruit	Freeze (juice and segments); use to make marmalade, jelly, cordial or candied peel	Also mix with other citrus fruits
lemon	Freeze (juice and slices); use to make marmalade, jelly, candy (slices, peel), cordial, curd, cheese or paste	Add juice to bland fruits to increase flavour
loganberry	Freeze; bottle; use to make jam, jelly or syrup	Mix with apples for jam and jelly
mulberry	Freeze; use to make jam or jelly	

Fruit	Method	Special points
orange	As for lemon	Use bitter oranges for marmalade and jelly.
peach	Dry; freeze; bottle; pickle; use to make jam, fruit butter or cheese, glacé fruit, brandied fruit or liqueur	Add lemon juice for flavour and pectin for jam
pear	Dry; freeze; bottle; pickle; dry store; use to make fruit butter or cheese, glacé fruit, brandied fruit or chutney	Add lemon juice for flavour and to protect against browning
pineapple	Freeze; bottle; use to make glacé fruit or brandied fruit	
plum	Dry; freeze; bottle; use to make jam, fruit butter or cheese, relish or pickle, or chutney	
quince	Bottle; use to make jam, jelly or fruit cheese and butter	
raspberry	Bottle; freeze; use to make jam, jelly, liqueur or syrup	
red currant	Bottle; freeze; use to make jam or jelly	Mix with fruits low in pectin
rhubarb	Bottle; freeze; use to make jam or chutney	Often mixed with other fruits for added flavour
strawberry	Bottle; freeze; use to make jam or syrup	Low in pectin. Often mixed with gooseberry or red currant for jam
tangerine	Freeze; use to make marmalade or cordial	Mix with other citrus fruit for marmalade
tomato	Bottle; freeze; pickle; use to make sauce, chutney (red and green) or relish	
whortleberry	See bilberry	

Choices of method for preserving herbs

All herbs can be frozen and dried. Mint can also be made into a jelly with apples and preserved in vinegar.

Choices of method for preserving meat, fish and game

Freezing, curing, smoking, drying or salting can all be successfully used, depending on the storage facilities available.

Pointers for shows

For shows or cookery examinations, take special care at all stages. Appearance is especially important.

Jams, jellies and marmalades

1. Be sure to use the right proportion of fruit rich in pectin to obtain a firm set (see p. 20).
2. Cook the fruit thoroughly, especially for marmalade, before adding sugar. The fruit or peel should disintegrate when rubbed between the fingers.
3. Check the setting point (see pp. 26-7) very carefully. Boil for the minimum time possible to retain the best flavour and colour.
4. Use preserving crystals (see p. 20) to help avoid scum. Remove all scum as soon as the setting point is reached. Draw a piece of kitchen paper over the surface after removing most of the scum with a draining spoon.
5. Ensure that any peel or fruit is evenly distributed, by cooling the jam or marmalade for about 10 minutes before potting. Jars must be warm, not hot; and if they are too cold, bubbles will form.
6. With jelly, avoid bubbles by gently pouring it down the inside of the jar when filling. Make sure the surface of the table is level before leaving the jelly to set.
7. Use standard-sized plain jam jars without screw lids. Fill the jar to the brim; then cover with a waxed disc and cellophane.
8. Give the jar a polish with a leather. Fix the label centrally between the seams of the jar and, likewise, half-way between the neck and the base.
9. The flavour should be fresh and free of off flavour such as metal.

Bottled fruit

1. Use jars specially designed for bottling. Clean them thoroughly, using a bottle brush.
2. Choose even-sized fruit for even cooking. If the fruit discolours easily, place it in salt water as it is prepared and rinse each piece just before packing. Peel carefully to keep the shape, and cut into neat, even-sized pieces to fit the jar. Pack, neatly, leaving as few spaces as possible.
3. For the syrup, use 50 g (2 oz) less sugar than the usual 250 g (8 oz) to each 1 litre (2 pints) of water (see p. 50); sugar tends to make the fruit rise in the jar.
4. Process only two jars at a time. Use the water bath method (see p. 51) but start with cold syrup in the jars and cold water in the pan. Slowly raise to the required temperature in 1½ hours. After 1 hour, it should be at about 54°C, 130°F. Process soft fruits and apple slices at 75°C, 165°F, for 10 minutes; stone fruits, citrus fruits, apple purée and soaked strawberries at 82°C, 180°F, for 15 minutes; and figs, tomatoes and pears at 88°C, 190°F, for 30 minutes (40 for solid-pack tomatoes).
5. Ideally, for exhibition, the fruit should be less than 6 mm (¼ in) above the base of the jar; there should be no gap at the top of the jar; the syrup must be clear; the metal ring should be free of rust and the top fixed on squarely; the jar should shine and the label should be centrally placed.

Pickles

1. Use clean, attractive jars with plastic or plastic-lined metal tops. Position the label centrally between the jar's two seams.
2. Position the fruit or vegetables attrac-

▲ An attractive appearance, as well as a good flavour, will help home-made preserves gain marks on the show bench.

ively in the jar, with the pieces of even size and shape. For mixed pickles, arrange the different fruits in a design or in layers of contrasting colour.

3. Check that the fruit or vegetable has not shrunk, and is covered to a depth of 1 cm ($\frac{1}{2}$ in) of vinegar.

4. The flavour of the vegetable or fruit should predominate over the spicy, acid or sweet flavours. With storage, the different flavours will blend.

5. The fruit or vegetable must be crisp to eat if raw, or tender, if cooked. It should not be tough, hard or mushy.

6. In straining the vinegar (see p. 57), place a white paper tissue in the nylon strainer. The vinegar should be brilliantly clear with no particles of spice in the bottom of the jar.

Chutneys

1. Use attractive airtight jars with plastic or plastic-lined metal tops. Give the jar a polish and position the label in the centre of the two seams of the jar, just below shoulder height, on a vertically flat surface.
2. Fill the jar to 1 cm ($\frac{1}{2}$ in) below the rim.
3. The chutney should be bright in colour.
4. The texture should be thick and even, with no free liquid. None of the ingredients should be hard, though some are sometimes deliberately left crisp.
5. Aim for a balance of sweetness, sharpness and spiciness. No one spice should predominate, and the chutney should not be too hot. The flavour of good chutney complements rather than dominates the accompanying food.
6. The flavour should be mellow; store the chutney for at least two months before exhibiting.

Jams, jellies and marmalades

It can be very satisfying to see a cupboard stocked with bright jars of red, green and gold, full of jams, jellies and marmalades. True, the supermarkets offer all these in abundance, but even the expensive varieties never quite achieve that "home-made" flavour. It's economical, too, to make preserves yourself, especially if you can spend time collecting wild fruits for free. Or you can follow the signs at the farm gates inviting you to "pick your own fruit". In this way, you can ensure that the fruit is really fresh—an essential when making jams and jellies. But the same preserves can also be made with dried, canned and frozen fruit; so we can emulate our frugal grandmothers, and their leisurely pace.

▼ A good-quality preserving pan is a worth-while investment; the wide shape allows for rapid boiling. Use a pressure cooker for tough fruits and a closely woven jelly bag for clear jellies. Weigh and measure accurately, and use a sugar thermometer to ascertain when the preserve will set. A jam funnel is useful for filling jars without spills.

What is jam?
Quite simply, cooked fruit boiled with sugar. The fruit must have the right proportions of acid and pectin in order to set properly, and enough sugar to make it keep. Marmalade is a preserve made with citrus fruits, and jelly is cooked, strained fruit juice boiled with sugar.

EQUIPMENT

The pan you will need A preserving pan needs to be wide, and fairly shallow, to allow for rapid boiling. It must have a capacity of at least 9 litres (19 pints). Thick aluminium or stainless

steel are the best materials. Brass and copper can also be used – but never iron or zinc.

Fruit should not be left in an aluminium pan, or the acid will begin pitting the metal.

Time can be saved by cooking the fruit in a pressure cooker.

Thermometer Use a sugar-boiling thermometer. The sugar concentration is correct when the preserve reaches 220°F. Keep the thermometer in a jug of boiling water; then dry it quickly before taking the reading. Keep the bulb above the base of the pan.

Jars Preserves soften when the gel is broken, so use jars only up to 450 g (1 lb) capacity.

Covers for jars The metal screw tops from bought jars are useful. Cover the jam with a thin layer of paraffin wax and promptly fit the top. Alternatively, use a waxed disc to cover the preserve, and a cellophane circle secured with a rubber band. Plastic skin as used for bottling (see p. 46) is useful for jars which need to be kept airtight during storage.

Jelly bag Bags made of closely woven fabric are available in shops selling hardware or wine-making equipment. Alternatively, one can easily be made from a square of linen or cotton, or any fabric that can be boiled.

Invert a stool or chair, securely tie the corners of the fabric to the four legs and place the bowl for collecting the juice underneath.

Scales A large size of scale pan is necessary.

Ladle For pouring jam into jars.

Measuring jug To hold up to 1 litre (2 pints).

Jam funnel This fits over the jars for filling.

INGREDIENTS

Fresh, slightly under-ripe, acid fruit makes the best preserve, since it ensures a good, firm set. Fruit can also be used that has been preserved by freezing, drying or canning; in fact, special cans of fruit are available for making jams and marmalades all the year round. Avoid using over-ripe fruit if possible; otherwise add extra fruit.

If, when the fruit is ripe, there is no time to use the whole supply, preserve it yourself by freezing or bottling. Pack the fruit in a freezer bag and attach a label identifying it and giving its weight and the recipe for which it is intended. If space is limited, cook the fruit first so that it takes up less room. Citrus fruits, for marmalade, can be individually wrapped in self-clinging plastic and packed in a freezer bag. They can also be bottled, either cooked to a pulp or with the peel finely sliced for marmalade. Frozen fruit should be cooked in the frozen state to help retain colour. When using preserved fruit for jams, jellies and marmalade, increase its quantity by an extra 10 per cent to ensure a good set.

Fruit for jam

The best fruits contain both acid and pectin in quantity.

Class 1 fruits For a good set use apples, apricots, black and red currants, damsons, green gooseberries and hard, sharp varieties of plum.

Class 2 fruits A medium set can be achieved using apricots, blackberries, loganberries, raspberries and the sweet, juicy types of plums.

Class 3 fruits These have poor setting quality. They include bilberries, elderberries, grapes, rhubarb, strawberries, pears, pineapple and cherries.

Well-set jam can be made with these fruits when a proportion of Class 1 fruits is mixed in. For example, red currants or gooseberries both blend well with strawberries.

Fruit for jelly

A large quantity of fruit is required for jelly. It is therefor all the more economical to use wild fruits, such as crab apples, blackberries and quince. A base of apple or apple skins helps give a good set; and jellies made from raspberries, gooseberries and loganberries are often mixed with apples for this reason. Black and red currants make excellent jelly and, for economy, can also be mixed with apple.

Fruit for marmalade

Citrus fruit – oranges, lemons, limes, tangerines and grapefruit – are used on their own or mixed together. The tangiest marmalade is made from bitter or sour oranges from Seville, Malaga or Sicily. Seville oranges give the finest flavour. They make a brighter preserve than sweet oranges because the peel becomes translucent when boiled with sugar.

Jelly marmalade needs a high proportion of fruit for a good set. The juice is strained and fine shreds of cooked peel added for decoration.

Choice of sugar

Special preserving sugar has crystals that dissolve easily and prevent the formation of scum; but granulated sugar can be substituted very successfully. Dark, moist, brown sugar and black treacle or molasses can be used to add flavour, especially to dark, "mature" marmalade. Substitute up to one quarter of the white sugar quantity and add an extra 50 g (2 oz) for each 400 g (1 lb) of sugar. Enough must be used to give a 60-per-cent sugar content to prevent growth of moulds and yeasts. If you make 3 kg (6 lb) of sugar it should make 5 kg (10 lb) of jam.

▶ Sharp, acid fruits make the best preserves. They may be fresh, frozen, dried or canned. Apples, gooseberries, black currants and plums make successful jam because they contain plenty of pectin, the setting agent. Strawberries and raspberries need mixing with gooseberries or red currants. Dried apricots are useful for making jam in winter, which is also the time to buy Seville oranges for making marmalade. Use preserving sugar or brown sugar.

BASIC METHOD FOR MAKING JAM

The aim is to cook the fruit thoroughly and then to add sugar and boil, to achieve a set as quickly as possible.

Preparing and cooking the fruit

Wash the fruit. Remove any damaged parts. Scrub citrus fruits; remove stalks, peels and stones; and cut the fruit up. Place in a pan with the measured amount of water and cook until pulpy. Lemon juice or citric acid may be needed at this stage to improve acidity, flavour and set.

Straining for jelly

Pour boiling water through a jelly bag. Leave to drip. The cooked fruit is then poured through the jelly bag and left for 30 minutes. Re-boil pulp with more water; strain. Mix the juices, test for pectin, and measure. Calculate the sugar needed.

Will it set? Testing for pectin

If the fruit is over-ripe or is known to contain little pectin, it is worth doing this test. Place 1 teaspoon of cooked juice or pulp in a glass, and add 3 teaspoons of methylated spirits. Stir well, and leave for about 1 minute. Pour into another glass. If plenty of pectin is present, one lump will have formed; if a moderate amount, the clot will break into three or four pieces; if insufficient, clots will only barely form. If necessary, add pectin stock or bought pectin.

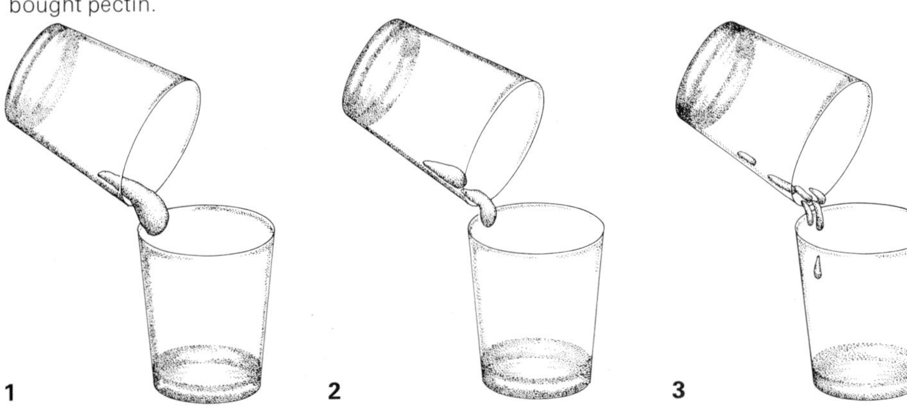

Adding sugar and testing for set

Add the calculated amount of sugar, depending on the amount of pectin in the fruit. Stir until dissolved; then bring to the boil. Boil on full heat until the mixture sets. Remove the pan from the heat and test its contents with a thermometer or a wooden spoon or by placing a little on a plate. If the preserve has set, a skin will have formed on the surface (see p.27).

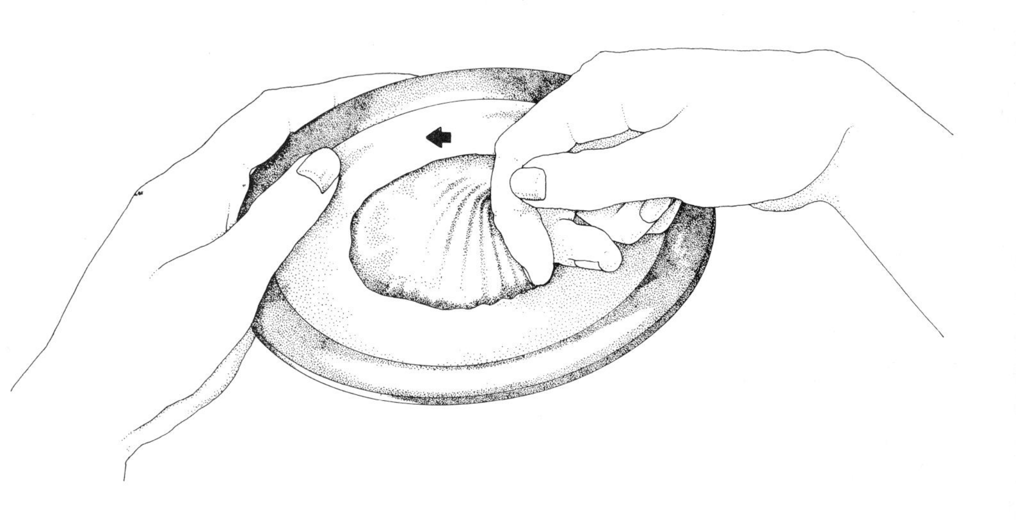

Finishing the preserve

Skim; leave to cool for 10 minutes to mix the jelly and the fruit pieces; and ladle into hot jars. Cover (see p.27); wipe the jar; and attach a label.

Pointers to perfect preserves Use fresh, under-ripe, acid fruit. Follow the recipe accurately to make sure that pectin, acid and sugar are present in correct proportions. Simmer the fruit until it is soft and broken down and its volume is reduced by one third. Strain the jelly without squeezing the bag, taking not more than a day.

Boil on full heat after dissolving the sugar (make sure the pan is large enough). Test for set after 5 minutes for small quantities (see pp. 26-7) and 10 minutes for large amounts. Continue testing at 5-minute intervals.

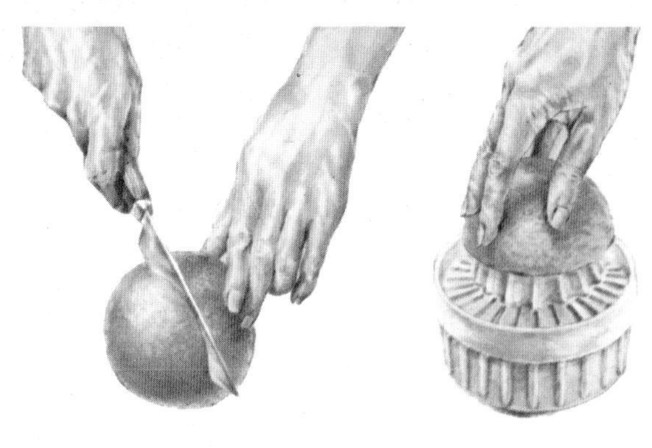
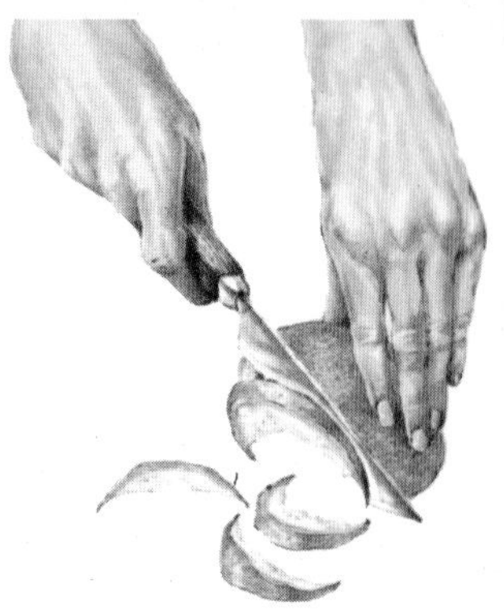

▲ Preparing oranges for chunky marmalade: squeeze juice and retain pips, stack about four peels together and shred with a stainless steel knife.

PREPARING FRUIT FOR JAM

Discard squashy, over-ripe or mouldy fruits. Cut out diseased or damaged parts. Remove stones from the freestone species of some fruits. Others, like damsons and cherries, are easiest left until their stones can be skimmed off the surface of the jam when they rise after the sugar has been added. Avoid washing soft fruits because it is difficult to drain them. To prevent sticking and minimize scum, rub the base of the preserving pan with buttered paper. Cook the fruit slowly with the measured amount of water and acid. Generally, the volume should decrease by a third.

PREPARING FRUIT FOR MARMALADE

Remove any stalks and scrub the fruit with a soft brush.

Chunky marmalade Squeeze the juice, then cut up the peel. Place in a preserving pan with the juice, water, and pips tied in a piece of muslin. The fruit can also be cooked whole in a casserole or covered saucepan. This is advisable for frozen fruit or when using a solid-fuel cooker with a slow oven. Cook until the peel disintegrates when pressed between the fingers. Remove the pulp and pips, return these to the pan, and cook for a further 30 minutes with the acid added. Meanwhile roughly cut up the peel with a knife and fork. Strain the pulp, press lightly through a nylon sieve into the preserving pan, and add the peel and sugar.

Thick marmalade Peel the fruit. Cook the shredded peel in a preserving pan with half the water. Some pith can be removed, if desired. Cut this up finely and put in a saucepan with the remaining water and fruit including pips. Cover; cook slowly for $1\frac{1}{2}$ hours; strain; and press lightly through a nylon sieve into the preserving pan with the peel.

Jelly marmalade Pare half the peel from the fruit. Shred very finely. Place in a saucepan with one third of the water. Cover and cook until tender, about 1 hour. Cut up the rest of the fruit and place in a large pan with the remaining water. Add the acid cover, and cook for about 2 hours. Add the strained liquid left from boiling the shredded peel; then pour through a scalded jelly bag and leave to drip.

Using a pressure cooker Reduce the water by half, using $\frac{1}{2}$ litre (1 pint) to each $\frac{1}{2}$ kg (1 lb) of fruit. The above methods can all be used; but it is best to cook fruit whole as for chunky marmalade to make a quick marmalade. Cook at high (15-lb) pressure for 20 minutes. For a large quantity, use a preserving pan for boiling with the sugar.

PREPARING FRUIT FOR JELLY

Only the juice is required, and accordingly much

lengthy preparation can be omitted. For example, there is no need to peel apples, top and tail gooseberries or remove the stalks of currants. Wash the fruit thoroughly and place in a pan. Hard-skinned fruits should be covered with water to soften them. Juicy fruits like raspberries and blackberries need very little water. Cook slowly for about an hour.

Straining fruit for jelly
Pour the cooked fruit into a scalded jelly bag and leave to drain for about 30 minutes. If the fruit is rich in pectin (see p. 20), tip the pulp back into the preserving pan after only 15 minutes. Mix with half the amount of water used for the first extraction and cook again for 30 minutes. Pour back into the jelly bag and leave to drip into the same bowl to mix the two extractions. If the pulp is left to drip overnight, make the jelly early next morning. The pectin content diminishes with standing, so make the jelly as quickly as possible. For a clear jelly, avoid the temptation to squeeze the jelly bag.

WHAT MAKES THE PRESERVE SET?

The all-important ingredient: pectin. It exists, in varying amounts, in most fruits: for example it forms the gummy substance sometimes seen on plums. Pectin combines with sugar to give the gel associated with jams and jellies.

Acid fruits make a good set because the acid helps extract the pectin. Fruits lacking in acid, such as strawberries, blackberries and cherries, need extra acid in the form of lemon juice or citric or tartaric acid. Lemon juice is the most complementary acid, but it is also the most expensive, though proprietary makes of lemon juice can be used. Both citric and tartaric acids are inexpensive and can be bought from the chemist or shops supplying wine-making equipment. One average-sized lemon contains about 30 ml (2 tablespoons) of juice. This is sufficient for $1\frac{1}{2}$ kg (3 lb) of fruit. Alternatively, add 2·5 ml ($\frac{1}{2}$ level teaspoon) of citric or tartaric acid. Cook the fruit with the acid before testing for pectin (see p. 22). Certain fruits, such as strawberries, cherries and marrows, must have pectin added in order to set.

Adding pectin
This may be done in one of three ways:

1. Make a pectin- and acid-rich stock from apples, gooseberries or red currants by covering with water then, cooking and straining the fruit as in jelly-making. If appearance is not important, sieve the fruit. Use about 250 g (8 oz) of pectin-rich fruit to each $1\frac{1}{2}$ kg (3 lb) of fruit lacking in pectin.

To save time, one can make up the stock or pulp in advance. Pour boiling stock or pulp into hot bottling jars; seal; then sterilize in a pan of boiling water for 5 minutes or in a pressure cooker for 3 minutes on low (5-lb) pressure. Use $\frac{1}{2}$ litre ($\frac{3}{4}$ pint) of pectin stock to each 3 kg (6 lb) of fruit.

Pectin stock can be added to a preserve that fails to set after boiling with sugar. Reboil 5 minutes, then test.

2 Mix in some fruit rich in pectin. For example: add gooseberries, red currants or lemons to strawberries and rhubarb.

3. Add commercial pectin. Bottled pectin can be bought quite cheaply. It is usually made from apples or citrus fruits. Follow the manufacturer's directions carefully.

ADDING SUGAR

The amount of sugar needed depends on the proportion of pectin in the fruit. The quantity of sugar given in a recipe will be the average amount for the variety of fruit. If the fruit is over-ripe, it will contain less pectin, and less sugar will be needed.

To keep well, a jam must have 60 per cent sugar to fruit. Each 5 kg (10 lb) of jam should contain 3 kg (6 lb) of sugar. If the sugar exceeds this ratio, it may crystallize. The required amount of sugar can be roughly calculated as follows:

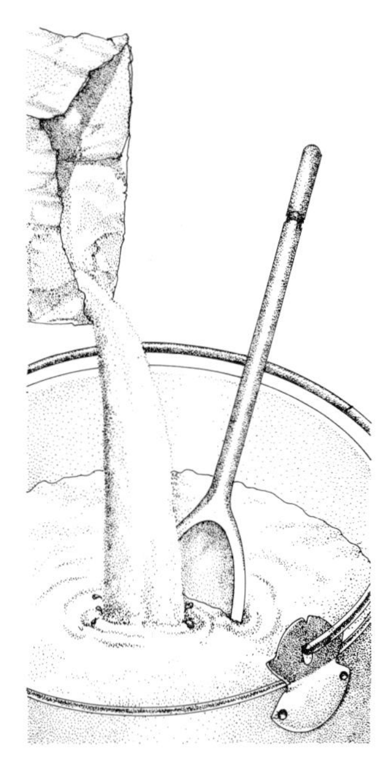

For jams and marmalades

Pectin-rich fruit: ½ kg (1 lb will set with 600 g (1 lb 5 oz) of sugar.

Medium pectin-rich fruit: ½ kg (1 lb) will set with ½ kg (1 lb) of sugar.

Low-pectin fruit: add extra pectin.

For fruit jellies

Pectin-rich juice: 550 ml (1 pint) of juice will set with 550 g (1¼ lb) of sugar.

Medium pectin-rich juice: 550 ml (1 pint) of juice will set with 350 g (12 oz) of sugar.

Low-pectin juice: add pectin stock or reduce by boiling.

Boil the cooked fruit or fruit juice, add pectin stock if necessary and stir in the sugar to dissolve it quickly. For a small quantity, heat the sugar in the oven on a plate. If sugar is added before the fruit is really soft, the fruit skins will toughen. It is especially important to avoid this in making marmalade.

BOILING THE PRESERVE

Boil on a full heat, timing until ready for testing. The mixture will foam and rise in the pan, which is why a large open pan is necessary. As a rough guide, when the foaming subsides and scum forms at the sides of the pan, the preserve is ready for testing. Boiling time depends on the quantity being made. The sugar concentration must be right. For a small amount, using 1½ kg (3 lb) of sugar, test after 5 minutes. For double this quantity test after 10 minutes. There are various ways of testing. Always first remove the pan from the heat.

1. Temperature test This is the most accurate method, but it requires a sugar-boiling thermometer registering at least 230°F.

Place the thermometer in a jug of boiling water. Stir the preserve; then dry the thermometer and take the temperature in the centre of the pan. Make sure that the thermometer bulb does not touch the bottom of the pan. The preserve will set when it reaches 212°F.

2. Flake test If a thermometer is not available this is the most accurate test for jellies. Dip a wooden spoon into the preserve. Turn the spoon while the preserve is cooling and setting on it. If the mixture sticks to the edge of the spoon in a large flake, the preserve is ready.

3. Plate test Chill a small plate. Place about 1 tea-

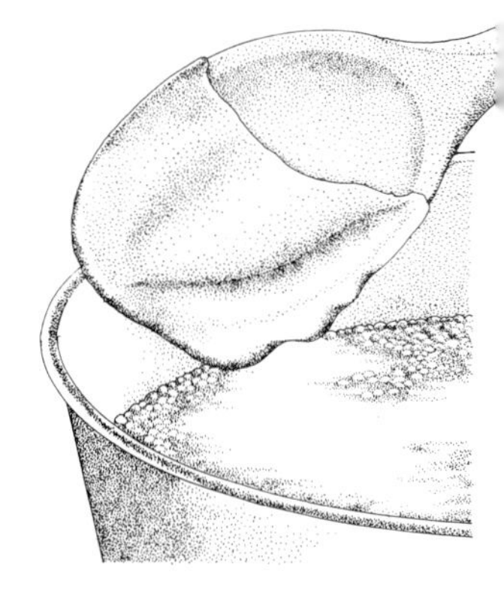

▲ Dip a wooden spoon in the preserve and turn it to cool. The mixture is ready if it forms large flakes on the spoon.

spoon of preserve on the plate and leave until cold. Gently push the jam or jelly with a finger; if the preserve has set, a skin will have formed on the surface which wrinkles when pushed.

FINISHING THE PRESERVE

Before starting to cook, collect sufficient jars and wash them thoroughly. Rinse them in hot water; then invert to drain. Place the jars in an oven on the lowest setting to become dry and warm. Wash any screw tops, or collect together waxed discs and cellophane tops. The discs can be cut from waxed paper or from cereal-box liners. Alternatively, to make an airtight cover, especially important for low- or non-sugar preserves (see p. 20), melt paraffin wax and pour it over the surface. Make sure it forms a seal at the edges.

Skimming

Place the warmed jars on a board or a wad of newspaper. Skim the preserve to remove the scum; but don't throw the scum away; it is delicious in stewed apple or jam tarts. If the preserve contains pieces, leave to cool for 10 minutes, stirring occasionally to make sure it is set sufficiently to support the pieces, otherwise they will all float to the top. This is especially important with jelly marmalade. Jellies must be skimmed very carefully. Use a piece of kitchen paper to remove the last bits of scum.

Filling jars and covering

To prevent air bubbles, pour the jelly in down the side of the jar. Fill each jar to the brim to allow for shrinkage. Cover while hot with a waxed disc, cut to the exact size of the top, or gently spoon melted paraffin wax over the surface of the preserve. While the jar is still hot, clean the rim before covering it with dampened cellophane secured by a rubber band or fitting a screw top. Wipe the jar with a damp cloth to remove any sticky marks; then leave undisturbed until quite cold and set. This is particularly important for jellies. When the preserve is used, the wax can be removed, washed, then used again. Label the jar, giving its contents and date. To check whether the jam contains its 60 per cent sugar, count the number of jars to a given quantity of sugar: 5 kg (10 lb) of preserve should be made from 3 kg (6 lb) of sugar. If more than 5 kg has been made and the jam will not keep well.

Storage

A cool, dry, airy cupboard or larder is the ideal place. To preserve the colour, it must also be dark.

Make sure the preserve contains sufficient sugar and that its cover is as airtight as possible. In damp conditions, use bottling jars and seal the preserve as for bottling (see pp. 50-51).

After opening, it is as well to store jars in the refrigerator, to prevent fermentation.

Diabetic jams, jellies and marmalade

Sugar is not allowed in a diabetic diet. The set must be obtained, therefore, from other ingredients – usually a mixture of gelatine and commercial pectin. For sweetening, Sorbitol and saccharin are used. Sorbitol contains about 100 calories per 25 g (1 oz), which is unsuitable in weight-reducing diets.

These preserves do not keep well and must be covered with airtight closures such as those used for bottling, or with melted paraffin wax. Store in the refrigerator after opening.

Low-sugar preserves for slimmers

People on reducing diets should not eat preserves at all. But on a "maintenance" diet, low-sugar jams can be eaten in moderation. Choose only fruits high in pectin (see p. 20), and use 350 g (12 oz) of sugar to each $\frac{1}{2}$ kg (1 lb) of fruit. Test by the flake or plate method; and cover and store as for diabetic preserves.

Liqueur-flavoured preserves

At very little extra expense, an everyday jam or marmalade can be turned into a treat. As well as doing wonders for a trifle these preserves can also double as sauces to serve with ice cream or creamy desserts.

Follow the usual recipe for making the preserve; then stir the liqueur into the pan just before potting. Use about 30 ml (2 tablespoons) for each jar of standard size – usually ½ kg (1 lb). Try adding brandy or sherry to apricot jam, kirsch to raspberry and black-currant jam, rum to black-currant, or an orange-flavoured liqueur (Grand Marnier, Curaçao or Cointreau) to strawberry.

Marmalade is delicious with whisky, rum or brandy added. Replace up to 25 per cent of the white sugar with an equal weight plus 20 per cent of black treacle or molasses for a "mature" appearance and flavour.

Freezer preserves

When fruit is cooked with sugar, some of the freshness of its flavour is lost. An uncooked preserve can be made by mixing fruit and sugar until the juice flows, then adding commercial pectin to make a set. It will keep in a refrigerator for a few weeks or in a freezer for six months. Once the gel is broken, the set collapses quickly. Pack the preserve in small containers; use at one serving.

Mixed-fruit preserves

Mix fruits for economy or to add flavour or pectin.

apple:	lemon, spice, blackberry, cherry, quince
marrow:	ginger, lemon, blackberry, plum
cherry:	red currant, gooseberry, lemon, orange
rhubarb:	raspberry, orange, ginger, damson

Conserves

These are whole-fruit jams, the fruit being suspended in a thick syrup. The fruit is layered with an equal weight of sugar and left for 24 hours, to toughen it, so that it stays whole, and to extract the juice. It is then boiled with acid until thick or until it forms a gel. The best fruits for conserves are: apple, loganberry, marrow, melon, pear, raspberry, rhubarb and strawberry.

What went wrong?

Mould on surface
Insufficient boiling, giving low sugar concentration and too large a yield.
Jar not filled
Poor cover.
Damp or warm storage.
Remove mould, re-boil jam and use for cooking.

Fermenting
Insufficient boiling, giving low sugar concentration and too large a yield.
Warm storage.
Remedy as above.

Poor set
Insufficient boiling after adding sugar.
Insufficient boiling to extract pectin and reduce.
Insufficient acid.
Too much sugar in relation to pectin and acid.
Over-ripe fruit.

Poor colour
Over-boiling after addition of sugar.
Over-ripe fruit.
Not enough acid.

Sugar crystals
Over-boiling, giving a high sugar content. Long crystals indicate too much acid.

Preserve shrunken in jar
Poor cover.
Warm storage.

Air bubbles
Preserve potted too cool.

Tough peel or skins
Insufficient cooking before sugar added.

Fruit or peel at top
Jar or preserve too hot at potting.

Cloudy jelly
Juice pressed through jelly bag or bag squeezed.

Layer of syrup at top
Too much acid.

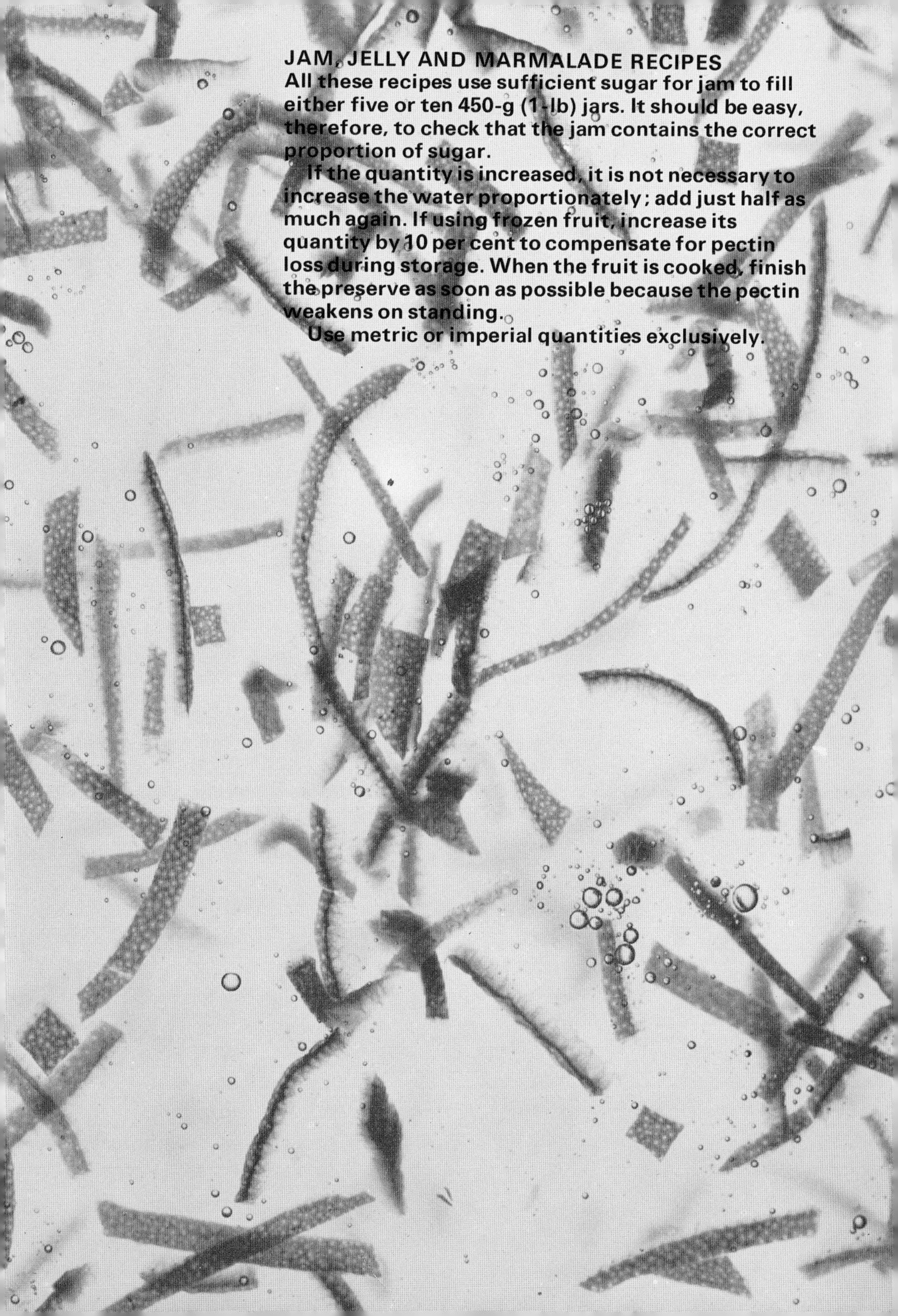

JAM, JELLY AND MARMALADE RECIPES

All these recipes use sufficient sugar for jam to fill either five or ten 450-g (1-lb) jars. It should be easy, therefore, to check that the jam contains the correct proportion of sugar.

If the quantity is increased, it is not necessary to increase the water proportionately; add just half as much again. If using frozen fruit, increase its quantity by 10 per cent to compensate for pectin loss during storage. When the fruit is cooked, finish the preserve as soon as possible because the pectin weakens on standing.

Use metric or imperial quantities exclusively.

APPLE

Apple jelly

Use well-flavoured apples or crab apples or add ginger or lemon. Cook with water just to cover; strain; re-boil after 30 minutes; and strain once more. Mix the juices, using 450 g (1 lb) of sugar to each 550 ml (1 pint) of juice.

Apple mint jelly

Make as above; cook a few sprigs of mint with the apple; then after adding sugar flavour the boiling jelly by holding a bruised sprig of mint in it. Finely chop some mint. After testing for set, add it to the jelly with green food colouring. Cool; stir; then pot in small jars.

Apple and elderberry jelly

Cook $1\frac{1}{2}$ kg (3 lb) of each fruit separately with water to cover. Use 350 g (12 oz) of sugar to each 550 ml (1 pint) of juice.

APRICOT

Fresh apricot jam

Use 3 kg (6 lb) of fruit to $\frac{1}{2}$ litre (1 pint) of water and 3 kg (6 lb) of sugar. Remove the stones; crack some and remove the kernels; blanch and skin these, and add to the fruit. Cook until reduced by one third; add sugar; boil until set; then pot.

Dried apricot jam

Use 1 kg (2 lb) of dried apricots, 3 kg (6 lb) of sugar, and the juice of 2 lemons or a 10-ml spoon (2 level teaspoons) of citric or tartaric acid. Soak the fruit for 24 hours in $3\frac{1}{2}$ litres (6 pints) of water. Cook for about 30 minutes with lemon juice or acid. Add the sugar; boil until set; then pot. Alternatively, omit soaking, reduce water by half and cook at high (15-lb) pressure for 20 minutes.

BLACKBERRY

Blackberry jam

Cook 3 kg (6 lb) of blackberries with 60 ml (4 tablespoons) of lemon juice, 5 ml (1 level teaspoon) of citric or tartaric acid or 250 ml ($\frac{1}{2}$ pint) of apple pectin stock. Add 3 kg (6 lb) of sugar and boil until set. Cool before potting. For seedless jams, press the fruit through a nylon sieve.

Blackberry and apple jam

Cook 2 kg (4 lb) of blackberries and 1 kg ($1\frac{1}{2}$ lb) of prepared apples in 150 ml ($\frac{1}{4}$ pint) of water each. Sieve the blackberries if desired; combine with the apple pulp; and add 3 kg (6 lb) of sugar. Boil until set; then pot.

Blackberry and elderberry jam

Prepare as for blackberry jam, but replace half the fruit with elderberries.

Blackberry jelly

Cook 4 kg (8 lb) of fruit in $\frac{3}{4}$ litre ($1\frac{1}{2}$ pints) of water and a 10-ml spoon (2 level teaspoons) of citric or tartaric acid. Strain; and add 450 g (1 lb) of sugar to each 550 ml (1 pint) of juice.

Blackberry and apple jelly

Use any proportion to a total of 3 kg (6 lb). Cook in

1 litre (2 pints) of water, using sugar as above.

Diabetic blackberry and apple jam
Crush $\frac{1}{2}$ kg (1 lb) of blackberries in the hands. Peel, core and chop $\frac{3}{4}$ kg ($1\frac{1}{2}$ lb) of apples. Cook in 125 ml ($\frac{1}{4}$ pint) of water until tender. Weigh $\frac{1}{2}$ kg (1 lb) of cooked apples and juice and place in a preserving pan with the blackberries and 575 g ($1\frac{1}{4}$ lb) of Sorbitol powder. Bring to the boil, stirring; add a knob of butter; and boil for 1 minute. Remove from the heat and add 5 crushed saccharin tablets and half a bottle of Certo, to give pectin. Sprinkle 25 g (1 oz) of gelatine onto 125 ml ($\frac{1}{4}$ pint) of cold water. Leave 5 minutes; then stir into the pan. Stir and skim for 5 minutes; then pour into clean, hot jars and cover with paraffin wax. (Recipe supplied by the British Diabetic Association.)

BLACK CURRANT

Black-currant jam
Cook 2 kg (4 lb) of black currants in $1\frac{1}{2}$ litres (3 pints) of water. Add 3 kg (6 lb) of sugar and boil until set. For economy, mix with an equal weight of apples or rhubarb.

Black-currant jelly
Cook 2 kg (4 lb) of black currants in 1 litre (2 pints) of water; strain; leave for 30 minutes; then cook the pulp in $\frac{1}{2}$ litre (1 pint) of water for a second extract. Strain; mix the juices; and measure. Test for pectin and add sugar accordingly (see p. 26): 550 ml (1 pint) of juice will set $\frac{1}{2}$ kg ($1\frac{1}{4}$ lb) of sugar.

DAMSON

Damson jam
Cook $2\frac{1}{2}$ kg (5 lb) of fruit in 1 litre ($1\frac{3}{4}$ pints) of water. Add 3 kg (6 lb) of sugar; stir until dissolved; remove the pan from the heat; then skim off the stones with a draining spoon (they rise to the surface after sugar is added). Boil until set.

Damson jelly
Cook 3 kg (6 lb) of fruit in 1 litre (2 pints) of water for the first extraction (see black-currant jelly); for the second, cook in $\frac{1}{2}$ litre (1 pint).

GOOSEBERRY

Gooseberry jam
Cook 2 kg (4 lb) of gooseberries in 1 litre ($1\frac{1}{2}$ pints) of water until reduced by one third. Add 3 kg (6 lb) of sugar and boil until set. Use a brass or copper pan to retain the green colour. In equal proportions, add strawberries, red currants or rhubarb.

Gooseberry jelly
Prepare as for black-currant jelly.

JAPONICA

Japonica jelly
Cook 2 kg (4 lb) of sliced fruit in $2\frac{1}{2}$ litres (5 pints) of water with 60 ml (4 tablespoons) of lemon juice until pulped, about 1 hour. Strain the juice; measure; and to every 550 ml (1 pint) add 450 g (1 lb) of sugar.

LOGANBERRY

Loganberry jam
Cook 3 kg (6 lb) of logan-

berries until pulped; add 3 kg (6 lb) of sugar; and boil until set.

Loganberry jelly Prepare as for blackberry jelly.

Loganberry conserve
Layer 2 kg (4 lb) of fruit with 2 kg (4 lb) of sugar. Leave for 24 hours. Place in a pan; add 60 ml (4 tablespoons) of lemon juice; boil; and cook for 5 minutes. Pour back into the bowl and leave for 48 hours. Boil until set – about 15 minutes – in the preserving pan. Check for setting by the flake or plate test (see pp. 26-7). Cool before potting and make an airtight cover.

MARROW

Marrow and ginger jam

Prepare marrow: cut into 1-cm ($\frac{1}{2}$-in) cubes and place 3 kg (6 lb) of them in a large bowl. Sprinkle with 1 kg (2 lb) of sugar and leave overnight. Squeeze the juice from 3 large lemons and shred the peel and pith. Place in a saucepan with $1\frac{1}{2}$ litres (3 pints) of water. Bruise 75 g (3 oz) of root ginger with a hammer; place in a muslin bag; and add to the pan. Boil, cover, and cook for $1\frac{1}{2}$ hours. Remove the ginger; pour the liquid through a nylon strainer; and press out the juice. Place the marrow and liquid from the bowl in a preserving pan with the ginger. Cook slowly until the marrow is tender, about 30 minutes. Remove the ginger, and add the strained liquid and 2 kg (4 lb) of sugar. Boil until the jam is set.

PLUM

Plum jam

Remove the stones, if possible. Place 3 kg (6 lb) of plums in a preserving pan with 250 to 500 ml ($\frac{1}{4}$ to $\frac{1}{2}$ pint) of water, depending on the juiciness of the fruit. Cook until the skins are softened; add 3 kg (6 lb) of sugar; and if necessary skim off the stones. Boil until set.

Plum and apple jam

Replace with apple up to half the plums given above.

QUINCE

Quince jam

Peel, core and cube 2 kg (4 lb) of quinces. Cook, covered, in $1\frac{1}{2}$ to 2 litres (3 to 4 pints) of water until tender, about 30 minutes. Uncover, and boil for 10 minutes to reduce. Add the juice of 2 lemons and 3 kg (6 lb) of sugar. Boil until set.

Quince jelly

Make as for japonica jelly. If the fruit is under-ripe, make two extracts: using 2 litres (4 pints) of water for the first and 1 litre (2 pints) for the second.

RASPBERRY

Raspberry jam

Cook 3 kg (6 lb) of raspberries until they are tender. Add 3 kg (6 lb) of sugar and boil until set. For economy, mix in fruits rich in pectin such as apples, gooseberries, plums, loganberries or red currants. Alternatively, add pectin stock made by boiling 250 g (8 oz) of any of these fruits in a little water and lightly pressing out the juice. If pectin stock is used, reduce the weight of the raspberries by $\frac{1}{2}$ kg (1 lb).

Raspberry and red-currant jam

Slowly cook $1\frac{1}{2}$ kg (3 lb) of raspberries until pulped. Cook $1\frac{1}{2}$ kg (3 lb) of red currants in 250 ml ($\frac{1}{2}$ pint) of water. Mix together; add 3 kg (6 lb) of sugar; and boil until set.

Raspberry preserve

Lightly crush $\frac{3}{4}$ kg ($1\frac{1}{2}$ lb) of raspberries and place in a bowl with 1 kg (2 lb) of castor sugar. Leave for 1 hour, giving an occasional stir to dissolve the sugar. Add 30 ml (2 tablespoons) of lemon juice and half a bottle (100 ml, 4 fl oz) of liquid pectin. Stir thoroughly. Pack into small jars or plastic containers, leaving 1·25 cm ($\frac{1}{2}$ in) head-space if the preserve is to be frozen. Store in a refrigerator for up to four weeks or in a freezer for up to six months. Once the gel is broken the preserve softens and it is best to use it up at one meal.

Raspberry conserve

Make as for loganberry conserve.

RED CURRANT

Red-currant jelly

Make as for black-currant jelly. For serving with meats, a sharp, very stiff jelly is required. This is made by not adding water when cooking the fruit.

Rhubarb jam
Flavour and pectin must be added. To each 2 kg (4 lb) of rhubarb, add 1 kg (2 lb) of loganberries, raspberries or gooseberries, or 250 ml ($\frac{1}{2}$ pint) of pectin stock.

RHUBARB

Rhubarb and ginger jam
Cut 3 kg (6 lb) of prepared rhubarb into chunks and layer in a large bowl with 3 kg (6 lb) of sugar. Leave overnight; then place in a preserving pan with the juice of 4 lemons and two 15-ml spoons (1 rounded tablespoon) of ground ginger. Boil until set.

STRAWBERRY

Strawberry jam
Cook $\frac{1}{2}$ kg (1 lb) of red currants or gooseberries in $\frac{1}{2}$ litre (1 pint) of water for 20 minutes. Strain, and press out the juice. Place 3 kg (6 lb) of strawberries in a preserving pan and cook until the juice is reduced by half. Add the red-currant juice and 3 kg (6 lb) of sugar, and boil until set. Cool before potting.

Strawberry conserve
Make as for loganberry conserve.

Strawberry preserve
Make as for raspberry preserve.

MARMALADE RECIPES

Many combinations of citrus fruits can be used. A pressure cooker is useful; it shortens the cooking time by two-thirds. Reduce the water by half.

Seville orange marmalade
Use $1\frac{1}{2}$ kg (3 lb) of fruit, 3 litres (5 to 6 pints) of water, 3 kg (6 lb) of sugar, and a 10-ml spoon (2 level teaspoons) of citric or tartaric acid or the juice of 2 lemons.

Shredded marmalade
Squeeze the juice from the oranges, tie the pips in muslin and shred the peel finely. Place in a preserving pan with the juice, acid and water. Cook until the peel is tender, about 2 hours. Remove the pips, and squeeze the bag and discard. Add the sugar, and boil until set. Cool slightly before potting.

Thick marmalade
Peel the oranges, and cut off some white pith. Shred the peel and place in a preserving pan with the lemon juice or acid and half the water. Cook for about 2 hours, until the peel is tender. Cut up the fruit and pith; and cook in the remaining water for $1\frac{1}{2}$ hours. Strain into the preserving pan, and press out the juice from the fruit pulp with a wooden spoon. Add the sugar; boil until set; and allow to cool before potting.

Chunky marmalade
Cook the fruit whole with the acid and measured water, either in a casserole in a slow oven for 4 to 5 hours, in a saucepan for $2\frac{1}{2}$ to 3 hours, or in a pressure cooker at high pressure (15 lb) in half the quantity of water for 20 minutes. Frozen oranges, too, can be cooked by these methods. Remove the fruit, place on a plate, peel off the skin and return the centres of the oranges to the pan. Crush the fruit; then leave to cook for a further 30 minutes (5 minutes in a pressure cooker). Roughly chop up the peel, using a knife and fork. Strain the liquid from the fruit; press it out further with a wooden spoon, and place it in a preserving pan with the cut up peel and sugar. Boil until set. Cool slightly before potting.

Old English marmalade

Make as for chunky marmalade, adding 2 rounded tablespoons of black treacle or molasses with the sugar.

Lemon marmalade

Use $1\frac{1}{2}$ kg (3 lb) of lemons to 3 litres (5 to 6 pints) of water and 3 kg (6 lb) of sugar.

Three-fruit marmalade

Select $1\frac{1}{2}$ kg (3 lb) of fruit from a mixture of grapefruit, lemons and oranges (sweet or bitter). Cook in 3 litres (5 to 6 pints) of water and add 3 kg (6 lb) of sugar.

Flavoured marmalade

Add 30 ml (2 tablespoons) of whisky, rum, brandy, sherry or Grand Marnier for each jar just before potting.

Diabetic marmalade

Quarter 4 large oranges. Cook in 1 litre (2 pints) of water until tender, about 2 hours. Cut up the fruit, removing the pips. Return to the liquid; add $1\frac{1}{2}$ kg (3 lb) of Sorbitol powder and boil quickly for 5 minutes. For pectin, add a bottle of Certo, and stir well. Sprinkle 50 g ($1\frac{1}{2}$ oz) of gelatine on to half a cup of cold water; leave to soak 5 minutes; then stir into the marmalade. Ladle into jars, and seal with melted paraffin wax.

Jelly marmalade

Pare the rinds from half of 1 kg (2 lb) of Seville oranges, shred finely, and boil until tender. Cut up the rest of the fruit; add $1\frac{1}{2}$ litres ($2\frac{1}{2}$ pints) of water and the juice of 2 lemons or a 10-ml spoon (2 level teaspoons) of citric or tartaric acid; cover; and cook for 2 hours. Add the liquid from the shreds; then pour through a scalded jelly bag. Leave to drip for 15 minutes; return the pulp to the pan; add $\frac{1}{2}$ litre (1 pint) of water; cook for 20 minutes; then strain, and mix the juices. Test for pectin (see p. 22): if weak, reduce by boiling. Add $1\frac{1}{2}$ kg (3 lb) of sugar and the shreds. Boil until set. Cool slightly before potting.

Lemon jelly marmalade and **grapefruit marmalade** can be made by the same method.

To use a pressure cooker

Tie the shreds in a muslin bag and place in a pressure cooker. Add the cut up fruit, 1 litre ($1\frac{3}{4}$ pints) of water and the lemon juice or acid. Cook at high pressure (15 lb) for 20 minutes. Drain the bag of peel but do not squeeze; pour the liquid through the jelly bag; and leave for 30 minutes to 1 hour. If the juice amounts to more than 1 litre ($1\frac{3}{4}$ pints), reduce by boiling; if it is less, make a second extraction. Add sugar, and finish as above.

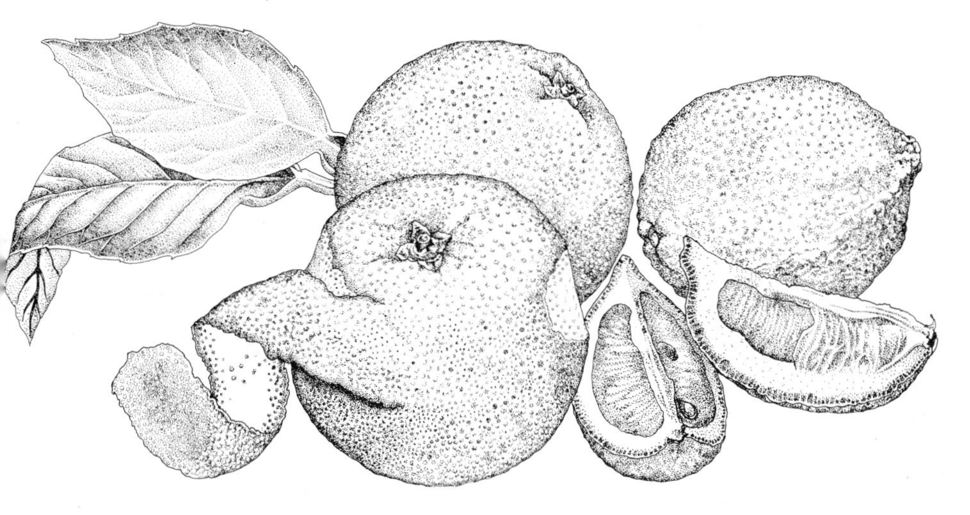

Fruit butters, cheeses and curds

These preserves use a lot of fruit and are best made when there is a glut. The fruits used for making jellies are also ideal for butters and cheeses, which is why these smooth thick preserves are often made together with jellies. This is an ideal arrangement, because the juice makes the jelly, and the pulp a butter or cheese. A butter is soft and spreadable; a cheese is thicker and is turned out of its container before serving; a curd is a rich preserve containing eggs and butter.

Fruits to use

Sharp fruits make the best butters, cheeses and curds. Use apples, blackberries, or both combined; black currants; damsons; quinces; or plums.

FRUIT BUTTERS AND CHEESES

1. Wash the fruit, and cut out any blemished or bruised parts. Leave the peel or stalks to be sieved out later.

2. Weigh the saucepan.

3. Chop or cut up any large fruits. Place the fruit in the saucepan, cover with water, and cook without a lid until pulpy. If it is to be used to make jelly, drain the mixture through a jelly bag.

4. Press the pulp into the pan through a nylon sieve, to remove all the pips, skin and fibre. Weigh the pan and its contents, and subtract the weight of the pan to arrive at that of the pulp.

5. Fruit butter Add 250 to 375 g (8 to 12 oz) of sugar to each $\frac{1}{2}$ kg (1 lb) of pulp. Add any spices. Stir until the sugar has dissolved; then cook for about 30 minutes, stirring occasionally, until thick. Test with a wooden spoon; if, when drawn over the pulp, it leaves an impression on the surface, the butter is ready. Pour the mixture into small jars, and, since it does not keep well, cover it with paraffin wax to make an airtight seal.

▲ A fruit cheese is thick enough when a spoon can draw a clean line through it.

Fruit cheese To each $\frac{1}{2}$ kg (1 lb) of pulp add $\frac{1}{2}$ kg (1 lb) of sugar, together with any spices. Cook for about 45 minutes, stirring occasionally, until there is no free liquid and the pulp is dry and solid. Test for consistency by drawing a spoon through the pulp: it should leave a clean line on the bottom of the pan. Pack the mixture into oiled wide-necked jars, individual soufflé dishes, or jelly moulds, and cover with waxed discs or paraffin wax. Leave two to three months.

6. Store both butters and cheeses in a cool, dark cupboard. Butters will keep for up to six months, cheeses for as much as two years.

FRUIT PASTES

Spread fruit cheese on to muslin or greaseproof paper on a wire cooling rack; and place in a very cool oven – or one that has just been switched off. When the cheese has dried out, cut it in squares or roll it into balls or lozenges, and coat these in granulated sugar.

Apple butter

Use crab apples or apples that have been used for jelly. Mix in blackberries, plums, oranges or lemons if desired. If apples are used on their own add spice.

2 kg (4 lb) cooking apples (windfalls will do)
1 litre (1$\frac{3}{4}$ pints) cider
5 ml (1 level teaspoon) ground cloves
5 ml (1 level teaspoon)

ground cinnamon
5 ml (1 level teaspoon) ground allspice
granulated sugar

Weigh the saucepan to be used. Wash the apples and cut out any bruised or damaged parts. Chop the fruit and place it in a saucepan with the cider. Cook, stirring occasionally, until it becomes pulpy. Sieve the apple mixture; return the purée to the saucepan; and add the spices. Weigh the pan and its contents. Add 350 g (12 oz) of sugar to each $\frac{1}{2}$ kg (1 lb) of purée. Cook, stirring, until thick enough for a wooden spoon to make an impression on the surface. Store in small jars covered with paraffin wax or other airtight seals.

Make gooseberry, blackberry, black-currant and damson, plum, and quince butters the same way; but substitute water for cider and omit spice. Add an orange to plum and blackberry butters.

Damson cheese

This cheese is delicious when dried to make a paste.

2 kg (4 lb) damsons
250 ml ($\frac{1}{2}$ pint) water
granulated sugar

Weigh the saucepan to be used. Wash the damsons; place them in the pan with water to cover, and cook until pulpy, stirring occasionally. Press through a sieve. Crack some of the stones, chop the kernels and add them to the purée. Weigh the purée in the pan, and add 1 kg (2 lb) of sugar to each 1 kg (2 lb) of purée. Cook, stirring frequently, until the mixture is thick enough for a spoon to leave a channel through it. Use plums, apricots, blackberries, black currants or mixtures of fruits. Instead of starting with fresh fruit, use the pulp left after straining for jelly.

FRUIT CURDS

Lemon is the classic ingredient; but oranges, apples, apricots, gooseberries or mixtures of fruits can also be used.

Lemon curd

4 medium-sized lemons
100 g (4 oz) butter
$\frac{1}{2}$kg (1 lb) sugar
4 eggs, beaten

Scrub the lemons; grate their rinds and squeeze out the juice. Place the juice and grated rind in a double saucepan over a low heat and add the butter and sugar. Cook, stirring, until the butter and sugar are melted. Add the eggs and cook, stirring, until the curd is thick enough to coat the back of the spoon. Pour into small jars, and seal with waxed discs or paraffin wax. Cover, and store in a cupboard for one month, or in a refrigerator for three months, or pack in small plastic containers and store in a freezer for up to nine months.

Orange curd

Make as above, but use 3 oranges and 1 lemon.

Apple curd

1 kg (2 lb) apples

$\frac{1}{2}$ kg (1 lb) sugar
2 eggs
250 g (8 oz) butter
5 ml (1 level teaspoon) ground cinnamon
5 ml (1 level teaspoon) ground cloves

Peel and core the apples and cook in a little water until pulpy. Add the other ingredients and cook over a moderate heat until thick. Take care not to boil the mixture. Pot, and cover with melted paraffin wax, or pour into small plastic pots. Cover, freeze, and store as for lemon curd. Use as a spread or a cake filling.

Apple and lemon curd

Make as above but replace the cinnamon and cloves with the juice and finely grated rind of 2 medium-sized lemons.

Syrups, cordials and juices

The fruit season, as the hottest part of the year, is also the time for cooling drinks. Home-made pure fruit juices, syrups and cordials are also valuable for topping desserts and ice cream or for making into fresh fruit jellies. Cordials are made from citrus fruits, and syrups from berry fruits. In winter the valuable Vitamin C contained in syrups and cordials makes them useful in the prevention of colds.

INGREDIENTS

Fruits Ripe, or even over-ripe, strongly flavoured, acid fruits are the most successful, notably black and red currants, damsons, strawberries, raspberries, loganberries, oranges, lemons, grapefruit and limes. Surplus tomatoes can also be made into juice. Free ingredients include rose hips.

▼ Use a double saucepan or a bowl over a shallow saucepan for cooking the fruit. A deep saucepan is needed for sterilizing bottles. An electric fruit press extracts the maximum amount of juice from hard fruits.

Pectin-destroying enzyme There are various brand names for this product, which can be bought from a chemist or a home wine-making supply shop. It is necessary only for the cold extraction of juice.

Campden tablets Available from the same shops as the pectin-destroying enzyme.

Citric acid Fruits lacking acidity require this extra ingredient.

EQUIPMENT

Saucepans A saucepan will be needed large enough to take a bowl. A deep saucepan is also essential, for sterilizing.

Stone jar For the no-cook method of extraction, or use a plastic bucket.

Fruit press The ideal, though expensive, device for hard fruits. A mincer or electric chopper is also necessary to break down the fruit before pressing.

Jelly bag To strain the juice, a piece of finely woven cotton sheeting can be tied to the legs of an upturned stool.

Squeezer This too is needed for citrus fruits.

Bottles Small bottles are the easiest to sterilize. The non-returnable 275-ml ($\frac{1}{2}$-pint) mixer bottles are ideal, with screw tops or corks.

BASIC METHOD FOR MAKING SYRUPS

The following process is used to make syrups from black and red currants, damsons, strawberries, raspberries, loganberries and blackberries. The aim is to break down the fruit and extract the juice without destroying its flavour. This is done by heat, by a pectin-destroying enzyme or by fermentation.

EXTRACTING THE JUICE

1. Heat Place the fruit in a large bowl and crush it with the hands or a piece of wood. Add about 250 ml ($\frac{1}{2}$ pint) of water to each $\frac{1}{2}$ kg (1 lb) of black currants and damsons and just a little water for berried fruits like raspberries, loganberries and blackberries. Cook the fruit slowly over a saucepan of boiling water, crushing occasionally with a wooden spoon. Tomatoes can be stewed.

2. Pectin-destroying enzyme Follow the directions on the pack; mix the enzyme into the crushed fruit; and leave in a warm place: overnight for soft fruits, and about three days for fruits with tough skins.

3. Fermentation By this method the wild yeasts on the fruit skins do the work of extracting the juice. Crush the fruit in a bowl or crock and leave in a warm place. Soft fruits should stand overnight; fruits with tough skins should be left for five days. If the fruit is left too long, it will make wine, not syrup, and the fresh fruit flavour will be spoilt. Strain off the juice as soon as bubbles appear on the surface.

4. Mechanical extraction Use an electric juice-extractor for most fruits or, for citrus fruits, a squeezer. A fruit press is useful for apples. Remove bad and bruised parts from the apples; then chop or mince them, or liquidize them with water containing a little Campden solution. Soak the fruit; then press it between cloths – cider cloths are the ideal if available.

If an electric juice-extractor is used, the fruit need not be soaked.

To press the juice use, preferably, a press or a juice extractor, or a jelly bag. Pour the juice into a scalded jelly bag and leave to drip overnight. Squeeze the bag between the hands to extract the maximum quantity. Some of the fruit solids can be removed first by pressing the pulp through a nylon strainer.

For a really clear syrup, pour the juice through a folded filter paper in a plastic funnel. If a small amount is being strained, a filter for making coffee with a shaped filter paper will do very well. Weigh the pulp and make into a fruit cheese (see p. 36).

ADDING SUGAR

Measure the juice and add 300 to 450 g ($\frac{3}{4}$ to 1 lb) of sugar to each $\frac{1}{2}$ litre (1 pint) of juice. Stir into the cold syrup until dissolved.

FILLING THE BOTTLES

Sterilize the bottles and tops or corks with Campden solution or by placing them in a saucepan of water, bringing to the boil and boiling for 15 minutes. Fill the bottles to within 3·75 cm ($1\frac{1}{2}$ in) of the top if corks are being used and 2·5 cm (1 in) in the case of screw tops. Fit the tops; and tie down the corks with wire or string to prevent them blowing off during sterilization.

STERILIZING

If bottles are to be kept for no more than four weeks,

they need only be stored in a refrigerator. Sterilizing, however, extends their storage life to one year.

1. Campden tablets Dissolve a crushed tablet in 15 ml (1 tablespoon) of boiled water and add to a ½ litre (1 pint) of syrup. Stir well; then bottle. The colour will fade slightly because of the bleaching effect of the solution. This method is easy, but there is a slight sulphur flavour. Store the bottles in a cool, dark cupboard and keep in the refrigerator when opened.

2. Heat This is similar to bottling. Place a rack or wad of newspaper on the base of a large deep saucepan. Place the bottles in the pan. Fill the pan with cold water, up to the level of the syrup in the bottles, using wads of newspaper to separate the bottles. Heat slowly to simmering point, and keep at this temperature for 20 minutes. Remove the bottles; firmly close their tops; then leave to cool. Dip the corks in melted paraffin wax to form a seal.

STORAGE

Store the bottles in a cool, dark cupboard, and keep in the refrigerator when opened.

Lemon cordial (see left)
8 large lemons
20 ml (2 rounded teaspoons) citric acid
1½ kg (3 lb) granulated sugar

Scrub the lemons, and pare the rind very finely, avoiding any white pith. A potato peeler or small sharp knife makes this job easy. Place the rind in a saucepan with 1 litre (1¾ pints) of water. Bring to the boil slowly; then remove from the heat and allow to cool. Halve the lemons; and squeeze out the juice and strain it through a piece of kitchen paper in a nylon sieve, into a large bowl or jug. Stir in the citric acid and granulated sugar. Strain the infusion from the lemon rinds through the sieve; then stir until the sugar is dissolved. Bottle and sterilize; use diluted.

Orange cordial

Orange rind is more pungent than lemon rind, so use the juice of 8 oranges and the rinds of only 4. Make as above, but add 25 g (1 oz) of citric acid and only 1¼ kg (2½ lb) of sugar.

Grapefruit cordial

Use the rind and juice of 3 large grapefruit and 1 lemon, ½ kg (1¼ lb) of granulated sugar, ½ litre (¾ pint) of water and 2×10-ml spoons (2 rounded teaspoons) of citric acid.

Rose hip syrup

Bring 1½ litres (3 pints) of water to the boil in a large saucepan. Mince 1 kg (2 lb) of rose hips and place immediately in boiling water. Bring to the boil, remove from the heat, and leave to infuse for 15 minutes. Strain through a scalded jelly bag and leave until almost dry; then return the pulp to the saucepan; add ¾ litre (1½ pints) of boiling water; and leave to infuse for 10 minutes. Strain; then mix the two juices, pour them back into the saucepan and boil rapidly until reduced by about half – the mixture should measure about 1 litre (1¾ pints). Add ½ kg (1 lb) of sugar and boil for 5 minutes; then bottle and sterilize.

Tomato juice

Chop up the tomatoes and cook until pulped. Rub through a fine sieve; and measure the purée. To each litre (2 pints) add 250 ml (½ pint) of water, 25 g (1 oz) of sugar, 10 ml (1 rounded teaspoon) of salt and 5 ml (1 level teaspoon) of citric acid.

Candied, crystallized and glacé fruits

Although these sweetmeats are considered luxuries, and are expensive to buy, they are really quite easy to make at home. A lot of patience is needed, but the results are well worth the effort, because making them at home costs so very little. The fruit is gradually impregnated with syrup, the strength being increased each day. They can be packed in individual paper cases and put into ornamental boxes to make presents.

INGREDIENTS

Fruits The best ingredients are small fleshy fruits such as plums and apricots; pineapple; cherries; and orange slices. Canned fruits, too, are often used. Angelica, orange and lemon peel and chestnuts can also be candied.

Sugar A mixture of granulated sugar and powdered glucose or dextrose gives the most penetrating syrup, but granulated sugar can also be successfully used.

BASIC METHOD FOR CANDYING FRUIT

1. Preparing the fruit

Fresh fruits: remove the stones from cherries, plums, greengages, peaches and apricots. Small plums like damsons can be pricked with a stainless steel or silver fork or a cocktail stick to allow the syrup to penetrate. Simmer the fruit very gently, to avoid breaking it; drain it and reserve 250 ml ($\frac{1}{2}$ pint) of the liquid.

Canned fruits: drain the fruit and measure the syrup. If necessary, make up the syrup to 250 ml ($\frac{1}{2}$ pint) with water.

2. Preparing the syrup

For fresh fruits, add 50 g (2 oz) of sugar and 100 g (4 oz) of powdered glucose or dextrose, or 150 g (6 oz) of sugar, to the reserved liquid. Stir over a low heat until dissolved.

For canned fruits, add 100 g (4 oz) of sugar and 100 g (4 oz) of powdered glucose or dextrose, or 200 g (8 oz) of sugar, to the reserved syrup from the can. Stir over a low heat until dissolved.

3. Candying process

Day 1 Place the fruit in a bowl; pour the boiling syrup over to immerse the fruit completely and leave overnight.

Day 2 Drain the fruit; dissolve 50 g (2 oz) of sugar in the syrup; bring to the boil; pour over the fruit; leave overnight.

Days 3 and 4 Repeat as for day 2.

For canned fruit:

Day 5 Drain the fruit; dissolve 75 g (3 oz) of sugar in the syrup; bring to the boil add the fruit to the pan and cook for 3 minutes. Pour into the bowl and leave for two days.

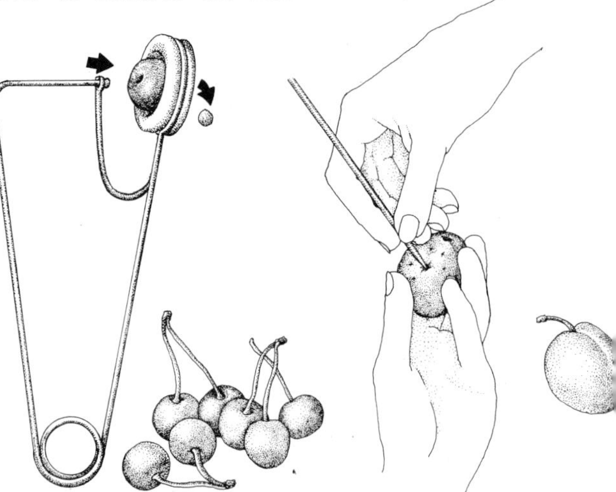

▲ A cherry stoner is a worthwhile gadget for keeping the fruit whole.

▲ Prick tough-skinned fruit with a cocktail stick in order for the syrup to penetrate them.

Day 7 Repeat as for day 5. The syrup should look like clear honey when cold. Leave for four days.
For fresh fruit:
Days 5, 6 and 7 Repeat as for day 2.
Day 8 As for day 5 for canned fruit.
Day 10 As for day 7 for canned fruit.

Leave the fruit in the syrup for up to three weeks. Finish just before it is required.

4. Draining
Place the fruit on a wire cooling rack with a plate below to catch any drips of syrup.

5. Drying
Place the cooling rack in the sun, or in the oven on its lowest setting or in the residual heat after baking. Leave to dry until the surface is no longer sticky.

6. Glazing the fruit
(a) In sugar: spear each piece on a skewer and quickly dip in boiling water; then coat in granulated sugar.
(b) In syrup: make a syrup by dissolving $\frac{1}{2}$ kg (1 lb) of sugar in 250 ml ($\frac{1}{4}$ pint) of water. Place a little in a teacup. Dip each piece of fruit in boiling water; shake off the excess; then dip the fruit in syrup. Place it on a wire cooling rack to drain; then dry the fruit in a very cool oven.

7. Packing and storing
Pack the fruit in boxes in layers with waxed paper in between. The box must not be airtight or the fruit may become mouldy. Place in petit-four cases to serve.

▲ To glaze fruit, dip each piece quickly in boiling water, then cover it thoroughly in granulated sugar.

RECIPES

Use metric or imperial quantities exclusively.

Candied peel

The peel from oranges, lemons and grapefruit can be candied for use in fruit cakes.
Day 1 Scrub the fruit well. Place the orange and lemon peel in a saucepan; cover it with water; bring to the boil; cover; and simmer for about 1 hour, until very tender. Test by pressing between the fingers. If cooked, the peel will disintegrate. Cook grapefruit peel the same way; but change the water two or three times during cooking. Add 50 g (2 oz) of sugar for the peel of each orange or lemon and 75 g (3 oz) for each grapefruit. Replace half the granulated sugar with powdered glucose or dextrose for the best appearance and longest storage. Stir until dissolved; then bring to the boil, remove from the heat and leave uncovered overnight.
Day 2 Bring to the boil and simmer for 5 minutes; leave uncovered overnight.
Day 3 Bring to the boil and simmer until the peel has absorbed most of the syrup. Place sections of peel separately on a wire rack, and fill the hollows with any remaining syrup. Leave to drain; then dry and glaze.

Marrons glacés

These crystallized chestnuts are luxury sweetmeats, but they are very easy to make.
Day 1 Use 1 kg (2 lb) of

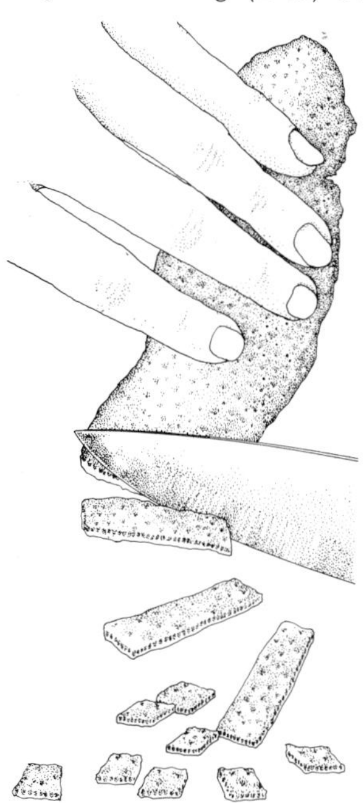

▲ Chop candied peel and store it in airtight containers, ready to use in fruit cakes and mincemeat.

sweet chestnuts. Nick the skins with a knife and cook a few at a time in boiling water for about 4 minutes each. Remove the shells and the brown skins.

Place the peeled chestnuts in cold water in a large saucepan, and bring slowly to the boil. Simmer very gently until tender, about 30 minutes, and drain. Use ½ kg (1 lb) of sugar and ½ kg (1 lb) of powdered glucose dissolved in 300 ml (a generous ½ pint) of water, for the syrup. Add the cooked chestnuts and bring to the boil. Remove from the heat and leave in the saucepan overnight.

Day 2 Bring the syrup to the boil; cover; and leave overnight.

Day 3 Add 5 ml (1 teaspoon) of vanilla essence to the syrup; bring to the boil; then leave to cool. Lift out the chestnuts carefully, to prevent breaking them. Place them on a wire rack with a plate underneath and leave overnight to drain. Press any pieces together.

Day 4 Give the nuts a glacé finish.

Pack the chestnuts in boxes between layers of waxed paper. If they are not to be used straight away, wrap them in foil. Place in petit-four cases to serve.

1. Nick the skin of each chestnut with a sharp knife before boiling.

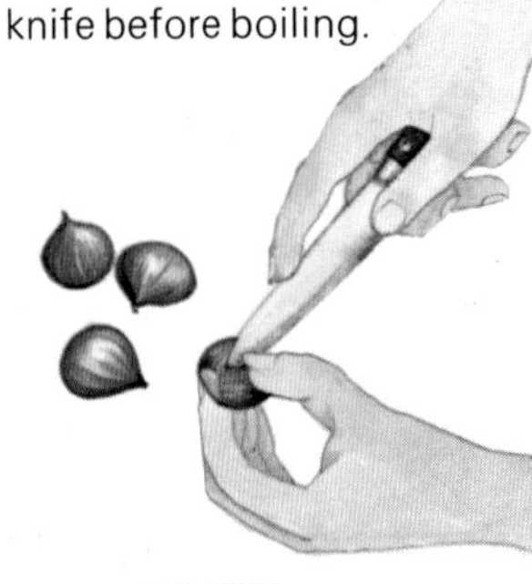

2. Cook 4 minutes.

3. Remove shells and brown skins of chestnuts.

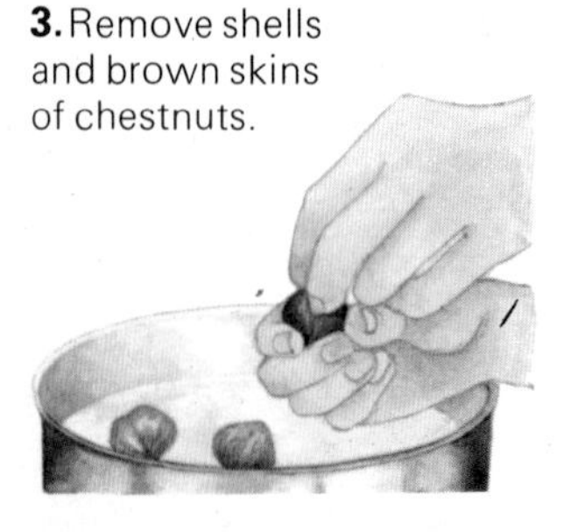

Mincemeat

This sweetmeat is a spiced mixture of fruits preserved by sugar and is a unique concoction of its own. No book on preservation would be complete without it, so here is a favourite recipe.

- ½ kg (1 lb) sultanas
- ½ kg (1 lb) currants
- ½ kg (1 lb) seeded raisins
- 125 g (4 oz) dried apricots
- 250 g (8 oz) mixed candied peel
- 1 large lemon
- 1 large orange
- 125 g (4 oz) blanched almonds
- ½ kg (1 lb) cooking apples, peeled and cored
- ½ kg (1 lb) beef suet
- ½ kg (1 lb) moist brown sugar
- 5 ml (1 level teaspoon) each of mixed spice, cinnamon and ginger
- 125 ml (¼ pint) whisky, rum or brandy (optional)

Place half the sultanas, currants and raisins in a bowl. Chop or mince the remainder and add. Chop or mince the apricots and the candied peel. Scrub the lemon and orange, grate their rinds and squeeze out the juice. Shred the almonds. Chop or grate the apples. Prepare and finely chop the suet. Add all to the bowl, and add the sugar, spices and spirits, if used. Mix thoroughly. Cover the bowl with a cloth and leave for two days. Stir well; then pot, and cover with waxed discs or paraffin wax, and a cellophane or screw top. Store in a cool, dry, dark cupboard. Leave the flavours to blend for at least one week before using.

Note: Modern homes without larders often have only rather warm kitchen cupboards available for storing preserves. Mincemeat might start to ferment in these conditions, especially if the spirits have been omitted. To prevent fermentation place the bowl in a very cool oven overnight or at 125°C, 250°F/Gas ¼, for 3 hours. Leave to cool, and then add spirits, if used.

▶ Make luxury sweetmeats from fresh or canned fruits at a fraction of the price of those in the shops. Pictured here are dark plums, citrus squares, tiny Wilkins oranges, candied peel, marrons glacés and glacé pineapple.

Bottling fruit and vegetables

For domestic purposes bottling is the equivalent process to canning. Most fruits can be successfully bottled: the fruit is cooked in syrup in a glass jar to sterilize it, then sealed to prevent contamination from moulds and micro-organisms. Fruits that need cooking are especially suitable for bottling. For vegetables, a high heat is needed to kill the organisms that cause spoilage in them, therefore a pressure cooker is essential.

EQUIPMENT

Saucepan Use a large saucepan or fish kettle, deep enough to immerse the upright jars in water completely.

Pressure cooker Check the manufacturer's handbook: the cooker must maintain a low (5-lb) pressure and be deep enough to take the jars. A high-domed cooker is necessary for 1-kg (2-lb) jars. Use only special bottling jars when processing in a pressure cooker.

Oven Place an oven thermometer on the centre shelf. Set the thermostat at 150°C, 300°F/Gas 2 and check the temperature after heating the oven for 30 minutes. Place jars at least 4·5 cm (2 in) apart on a wad of newspaper on a baking sheet.

Jars Special bottling jars of several kinds with metal or glass tops are available in ½- and 1-kg (1- and 2-lb) sizes. The initial cost is quite high, but the jars last for many years if handled carefully. Metal tops or rubber rings can be replaced every year quite cheaply. One type has a metal lid with a fixed rubber gasket and a lacquered screw ring to hold the lid in place. After processing, the vacuum holds the lid on the jar and the screw ring can be removed, washed, dried and stored in a plastic bag to avoid rusting. Another type of lid has a plastic gasket and clips at the side to hold the lid in position on jam jars. Some of these tops fit coffee jars. The same plastic is sold in sheet form. It can be cut to size for covering different shaped jars but it must be tied down securely.

Wash the jars well. Rinse in hot water, then invert and drain. The jars are easier to fill if they are slightly wet.

Pour boiling water over rubber rings to sterilize and soften them.

Thermometer Use for the water bath method. A thermometer is useful in finding the time needed to reach the necessary temperature. An ordinary one registering the boiling point of water (100°C, 212°F) is adequate. If the jars of fruit are being exhibited and appearance is important, use a thermometer to check that the temperature is being raised very slowly to simmering point (see p. 51).

BASIC METHOD FOR BOTTLING

1. Fill the jars with fruit, leaving as few air spaces as possible. Top up to the rim with syrup (brine for vegetables).

2. Close the jar; fit the top securely. Loosen screw rings one quarter turn.

3. Process jars in
(a) water in a deep pan.

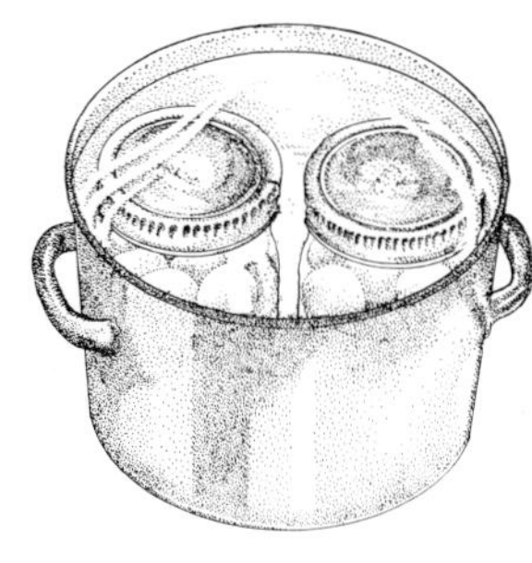

(b) an oven at 150°C, 300°F/Gas 2.

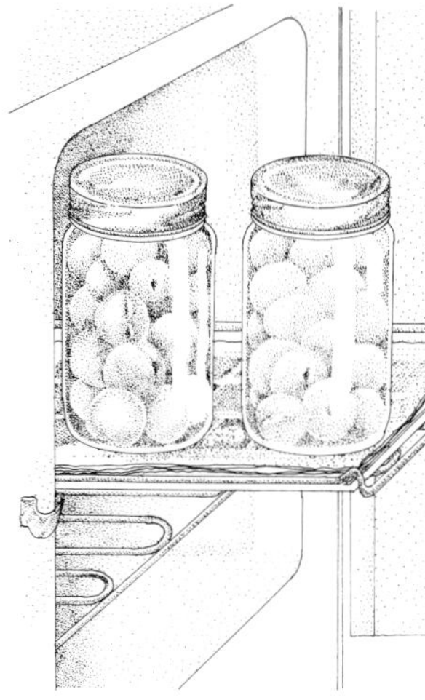

(c) a pressure cooker at low (5-lb) pressure. See chart on p.53.

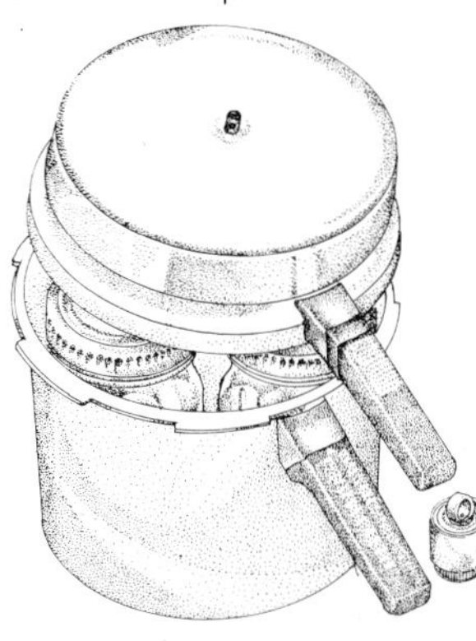

4. To remove the jars from the pan or pressure cooker, lift out with a pair of tongs.

5. To check that a vacuum has formed, test each jar by lifting it up by its lid.

PREPARATION OF FRUIT FOR BOTTLING

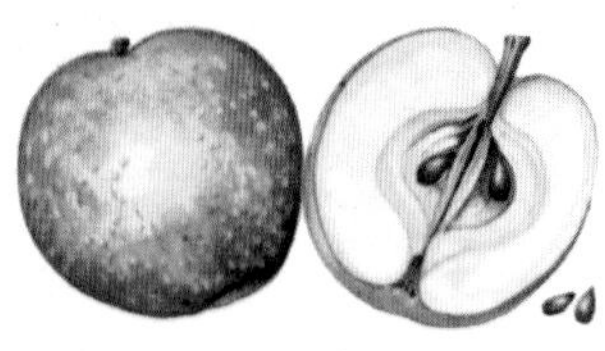

Apples Peel; core; slice; and leave in water containing 1 rounded teaspoon of salt to 1 litre (1¾ pints). Rinse before packing. For solid pack, leave the slices in a sieve in boiling water until pliable, about 3 minutes. If using windfall apples, make a purée.

Apricots Best halved. Cut round the line: twist then remove the stone; pack quickly to prevent browning.

Bilberries (blueberries, whortleberries, blaeberries) Remove the stalks. Pack into jars tightly, a third at a time, covering each layer with syrup.

Blackberries Wash only if necessary. Pack as for bilberries. If mixing with apples, scald these as described, for a solid pack.

Black currants As for bilberries.

Cherries Remove stones. For sweet cherries, add 1 level teaspoon of citric acid to each litre (1¾ pints) of syrup.

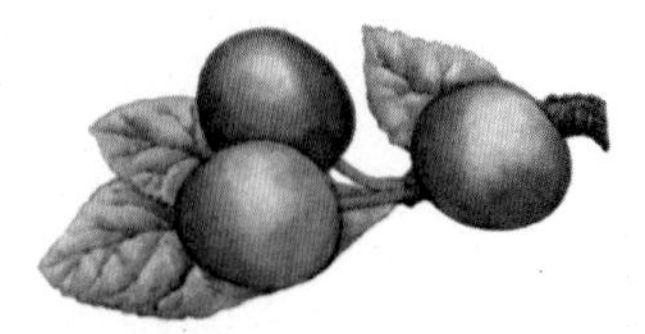

Damsons Remove stalks. Wash to remove bloom.

Figs Use ripe fruit. Remove stalks; peel if desired. Add 1 level teaspoon of citric acid per litre (1¾ pints) of syrup.

Gooseberries Choose small, green gooseberries. Top and tail, removing a small piece of skin to allow the syrup to penetrate.

Grapefruit Cut off peel. To remove segments, cut between pith; remove pips. Pack tightly into small jars.

Lemons Not worth bottling except as slices for drinks.

Loganberries Spread out to allow any maggots to be removed. Pack a third at a time with syrup, as for bilberries.

Oranges As for grapefruit.

Peaches Remove the skins by scalding for 1 minute and placing in cold water; then process as for apricots.

Pears Dessert pears are best. To prevent discoloration, place in acid salt water (1 rounded teaspoon of salt and 1 level teaspoon of citric acid to 1 litre or 1¾ pints of water). Poach cooking pears in syrup (100 g of sugar to ½ litre of water, or 4 oz to 1 pint) until tender.

Pineapples Core, and cut into segments or rings.

Plums If small, prepare as for damsons; bottle whole. Halve freestone varieties as for apricots; remove stones. Process quickly to prevent discoloration.

Quinces As for cooking pears. Mix with apples if desired.

Raspberries As for loganberries. Wash only if necessary.

Rhubarb Trim; then wipe. Cut to fit jars. Rhubarb can be packed raw, but will fit better if first covered with boiling syrup and left to soften for about 10 hours. Drain; pack; then cover with syrup reduced by boiling fast for 5 minutes. For exhibition, use soft or boiled water to prevent white specks forming.

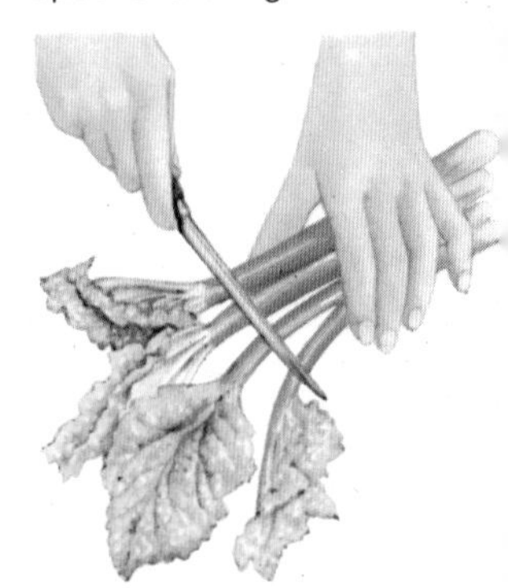

Strawberries Cover with boiling syrup and leave overnight. Drain well, then concentrate the syrup by boiling quickly for 10 minutes. Cover the fruit. Add red colouring to the syrup, if desired.

Tomatoes Best as a solid pack. Skin, cut into quarters and pack tightly into small jars. Sprinkle each layer with some salt and a little sugar. No water is required.

KILNER
Dual Purpose
JAR
KILNER
Dual Purpose
JAR

FILLING THE JARS

Pack the fruit tightly to avoid the chance of air bubbles. Aim for a high proportion of fruit to liquid because this liquid cannot always be used. Remove the stones or blanch the fruit (see *Preparation of fruit*) so that it can be tightly packed. Use a wooden spoon handle to position the fruit (particularly important if the jar is to be exhibited).

Pack fruit that discolours, such as apricots, in layers; then cover each layer with syrup. Pack soft fruit one third at a time cover with syrup and tap out any bubbles.

Some fruits, such as apples and tomatoes, are useful in purée form, especially if space for storage is limited and/or only small jars are available. Soft fruits make successful purées and may be served as sauces (see p. 67).

Syrup

Flavour and appearance of the fruit is improved by bottling in syrup made from sugar and water, though water alone is adequate and should be used for diabetics and slimmers. Sugar substitutes are best added when the fruit is eaten.

For sour fruits such as rhubarb and dessert fruits like peaches, dissolve 250 g (8 oz) of granulated sugar in each $\frac{1}{2}$ litre (1 pint) of water to make syrup for six $\frac{1}{2}$-kg (1-lb) jars. Boil for 1 minute.

For solid packs, use a stronger syrup, about $\frac{1}{2}$ kg (1 lb) of sugar to each $\frac{1}{2}$ litre (1 pint) of water, because less syrup can be added.

FLAVOURING THE SYRUP

Golden syrup or honey can replace sugar. Whole spices and orange or lemon rind can be boiled with the syrup. Liqueurs, brandy or wine can be added. The flavour of some bland fruits such as figs is improved by lemon juice or citric acid.

Most flavourings need straining through a sieve lined with muslin or a paper towel before adding to the jar.

For fruits which lose colour on cooking, like strawberries, add a little food colouring to the syrup. Fill the jar to the neck, twist, and tap lightly on the table to remove air bubbles. Place the jar on a plate to catch any spilt syrup, then fill to the brim with syrup.

CLOSING THE JARS

For the **water bath** and **pressure cooker methods**

▲ Bottle fruit in any type of jar. Cover tightly with preserving skin.

▲ After processing, the skin will shrink into the jar. Label the jar, giving the date of processing.

of processing, fill the jars with syrup, fit the tops on he jars and secure tightly. Make sure the rim of the jar s clean and no pieces of ruit overlap it to spoil the seal. Give the screw rings a quarter turn back to loosen slightly and to allow any air o bubble out during processing. For the oven method, place the top over he jar but do not secure.

PROCESSING THE JARS

he method to choose depends on the equipment vailable. The aim is to cook nd sterilize the fruit in the ar then close the top securely to avoid contamination. Very slow cooking is ecessary to allow heat to enetrate the centre of the ars and to preserve the appearance of the fruit. Cooking too quickly forces syrup ut of the jar, and the fruit ses from the bottom and is ot covered by syrup at the op. The **water bath method** is the easiest to ontrol and is the one to use, n a modified form, if the ars are to be exhibited (see . 16). For a large number f jars, choose the **pressure ooker method**, for speed, r the **oven method.** Large, ll jars become overcooked t the top in the oven before ne bottom is cooked.

Water bath method Place thick cloth, wad of newsaper or rack in the saucean to raise the jars above ne base. Fill the warm jars ith fruit and hot syrup. lose then loosen screw ngs. Place in the saucepan, allowing a 4·5-cm (2-in) space between each. Cover the jars with hand-hot water (50°C, 100°F). If this is not possible, cover with a tight-fitting lid. Bring to simmer-

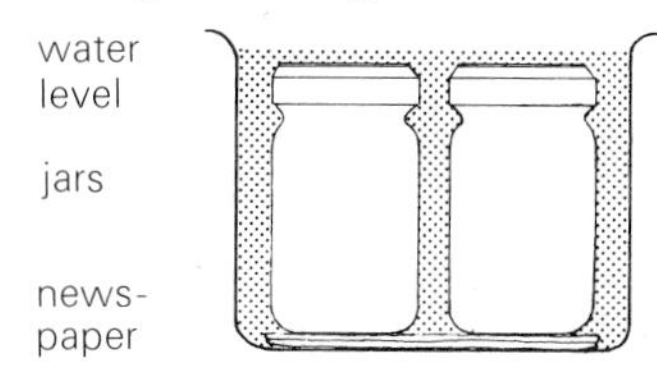

ing point (88°C, 190°F) over about 30 minutes. Maintain at this temperature for the time shown in the chart on p. 53. Large jars of over 1-kg (2-lb) capacity will take a longer time. Bale out the water with a cup to expose the tops. Place the jars on a board. Tighten the screw rings and leave the jars until cold, preferably overnight.

Pressure cooker method Use only special bottling jars. Jam jars may not withstand the pressure. Invert the trivet and place the empty jars in the cooker on the trivet; close the lid; and check that the jars do not block the safety valve or steam vent.

Top up the fruit in the jars with boiling syrup, close then loosen the screw rings,

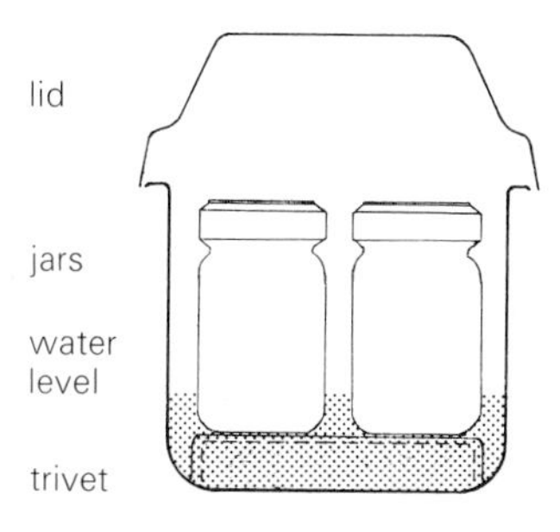

pour water to a 2·5-cm (1-in) depth into the cooker and bring to the boil. Place the jars on the upturned trivet, and close the lid. Place the cooker over a moderate heat, allow the steam to escape, if necessary, and bring up to low (5-lb) pressure. This should take 5 to 10 minutes. Maintain at this pressure for the time listed on the chart, p. 53. Fluctuation in pressure will force out the syrup. Remove from the heat. Leave for 10 minutes before removing the lid. Tighten the screw rings and leave to cool overnight.

Oven method Set oven to cool (150°C, 300°F/Gas 2). Fill the jars to the shoulder with boiling syrup, place a metal top over each and put on a wad of newspaper on a baking sheet. Leave 4·5 cm (2 in) between the jars and the side of the oven to allow the warm air to circulate. The processing time depends on the number of jars in the oven. Two 1-kg (2-lb) jars will take the same time as four ½-kg (1-lb) jars. Aim to seal the jars as quickly as possible after processing. Remove one at a time from the oven, quickly top up with boiling syrup, then secure the top. One advantage of this method is that if the fruit takes up less space after cooking, it can be topped up with cooked fruit.

TESTING FOR SEAL

Leave overnight. Remove screw rings and lift up each bottle by its lid. If the lid remains attached, a vacuum has been formed. If the lid comes off, drain off the syrup, check the jar and its seal for flaws, and reprocess. Jars covered with plastic skin will be airtight if the skin has been drawn downwards and forms a hollow. Follow the manufacturer's instructions carefully. If the fruit has slightly risen above the base of the jar, it is likely that a vacuum has been formed.

STORAGE OF JARS

Remove any screw rings. Wash the jars and rings in warm water to remove traces of syrup. Dry carefully and label the jars, noting the variety of fruit and the date. Details of bottling (e.g. in water or syrup, strength of syrup) may also be given.

Rub over the metal top and ring with a little oil to prevent rusting. Store rings separately in a plastic bag. It is easier to check the seal of the jars periodically if the metal ring is not attached. Store in a cool, dry, dark cupboard. To open the jars, prise off the lid by placing the point of a knife between seal and jar.

BOTTLED FRUIT FOR DIABETICS OR SLIMMERS

Use boiled water instead of syrup. To use the fruit, pour into a dish, sprinkle with Sorbitol, for diabetics, or low-calorie sweetening powder, for slimmers. Leave for 1 hour before serving.

BOTTLED FRUIT FOR JAM-MAKING

Cook the fruit following a jam recipe, using a measured amount of water. Pack the boiling pulp into jars, omit the syrup, cover, and process for 5 minutes. Cool, test and label, noting quantity and instructions for finishing.

BOTTLING FRUIT WITH PRESERVATION TABLETS

This method is useful when time is short. It involves packing the fruit into jars, covering the fruit with a solution of sulphur dioxide in cold water, then sealing the jar. No heat processing is required. Sulphur dioxide is available in the form of Campden tablets, from chemists or shops selling wine-making equipment. One tablet dissolved in 250 ml ($\frac{1}{2}$ pint) of water is sufficient to sterilize $\frac{1}{2}$ kg (1 lb) of prepared fruit.

The most suitable fruits to use are plums and other stone fruits and apples. Soft fruits, sweet cherries, pears, tomatoes and vegetables are not suitable for preserving by this method.

The chemical flavours the fruit, but boiling in an open pan before use will get rid of the sulphur flavour.

Chemical preserving method

Prepare the fruit, weigh it, and pack into clean glass or stone jars. Cover with a measured amount of sulphur dioxide solution made by crushing a Campden tablet and dissolving in 250 ml ($\frac{1}{2}$ pint) of cold water for each $\frac{1}{2}$ kg (1 lb) of prepared fruit. Cover the fruit with this solution; and close the jar with a glass top, a synthetic skin, or a layer of melted paraffin wax and a metal top. A piece of sheeting dipped in paraffin wax or mutton fat and tied over the top of

What went wrong?

Fruit discoloured at top of jar	Temperature changes in processing have spilled syrup over. Delay in processing. Jars not submerged. Lacquered lid scratched.
Unsealed jar	Fragment of fruit lodged under ring. Distorted lid or rubber band. Insufficient processing for a vacuum. Metal ring not tightened.
Fruit fermenting in sealed jar	Over-ripe fruit. *Discard.*
Fruit risen in jar	Over-ripe fruit. Too much sugar in syrup. Fruit loosely packed. Fruit overcooked. Temperature has risen too quickly.
Surface of fruit mouldy	Jars not submerged. Too many jars in oven at once. Under-processed.

PROCESSING TIMES FOR BOTTLING FRUIT

All times are shown in minutes

Fruit	**Water bath**	**Pressure cooker**	**Oven**	
berries and currants, apple slices	2	1	Up to 2 kg (4 lb)	40
			$2\frac{1}{2}$ to 5 kg (5 to 10 lb)	60
gooseberries, rhubarb, stone fruits (whole), citrus fruits	10	1	Up to 2 kg (4 lb)	50
			$2\frac{1}{2}$ to 5 kg (5 to 10 lb)	70
solid-pack apples,purée halved stone fruits, pineapple, strawberries	20	4	Up to 2 kg (4 lb)	60
			$2\frac{1}{2}$ to 5 kg (5 to 10 lb)	80
whole tomatoes, figs, pears	40	5	Up to 2 kg (4 lb)	70
			$2\frac{1}{2}$ to 5 kg (5 to 10 lb)	90
solid-pack tomatoes	50	15	Up to 2 kg (4 lb)	80
			$2\frac{1}{2}$ to 5 kg (5 to 10 lb)	100

the jar whilst still pliable is also suitable. Store the jars in a cool place. Red fruits will lose their colour, but it will return when the fruit is boiled before use. Sweeten to taste just before serving.

BOTTLING VEGETABLES

Vegetables are more difficult to preserve by bottling than fruits. Because they lack acidity, vegetables become contaminated by bacteria from the soil, which are very difficult to destroy. The high temperature necessary for bottling vegetables can only be achieved by using a pressure cooker, which will give a medium (10-lb) pressure. To make sure the heat penetrates the vegetables sufficiently use small, $\frac{1}{2}$-kg (1-lb) jars, of the type specially made for bottling.

Preparing the brine

In place of the syrup used in fruit bottling, a salt solution is used to cover vegetables. Use about 25 g (1 oz) of kitchen salt (not iodized) to each litre (2 pints) of water. Green food colouring can be added to peas and beans, and lemon juice to white vegetables, to preserve a natural colour.

▼ Sort asparagus into stems of even thickness; scrape their bases; and trim them to the length of the jar in which they are to be stored.

Preparing the vegetables

See the chart. Wash root vegetables thoroughly. All vegetables should be blanched before packing. Place in boiling water, return to the boil and time as in the chart.

Preparing and filling the jars

Follow the instructions for fruit (see p. 50). Use small-size bottling jars with screw rings or clips. Top up with boiling brine.

Processing the jars

Pour boiling water into the pressure cooker to a depth of 5 cm (2 in). Invert the trivet and place in the cooker. Loosen screw rings a quarter turn and place in the cooker at least 5 cm (2 in) apart. Close the cooker, and bring to the boil. Steam for 5 minutes before placing pressure weight in position or closing the vent. Bring up to medium (10-lb) pressure and maintain for time indicated on the chart. Keep a check on the pressure to make sure it does not fluctuate. Leave to cool slowly for 10 minutes before opening, then lift out the jars and tighten the screw rings. Leave to cool.

Testing for seal

Check the bottles by holding by the top, as when bottling fruit. Use up any vegetables from the bottles that have not sealed: they would be over-cooked and would deteriorate in flavour if processed a second time.

The high temperature and long processing forces some of the brine out of the jars, but this does not affect the keeping qualities.

Storage of bottled vegetables

If in doubt about the quality of the vegetables at any time during storage, discard

BOTTLING VEGETABLES

Vegetable	Preparation	Blanching time (in minutes)	Processing time (in minutes)
asparagus	Sort into stems of even thickness; scrape bases; trim to fit jar	2	30
beans	broad: Sort into sizes	3	35
	French: Cut off ends; leave whole or cut in chunks	3	
	runner: Remove strings and ends; cut in chunks	3	
carrots	Choose small carrots, cut off tops, blanch for 1 minute, rub off skins. Cut into rings or sticks	5	35
celery	Wash and trim to height of jar. Cut heart lengthways into four. Add lemon juice to blanching water	5	30
mushrooms	Wash well, trim stalks	5	35
new potatoes	Choose small potatoes; wash; boil for 5 minutes; then rub off skins	5	40
peas	Grade for size. Add green food colouring and sugar (1 level tablespoon to $\frac{1}{2}$ litre or 1 pint)	1	40

them without tasting. Vegetables can be very dangerous if not processed properly. Check each jar before serving and if there is any uncertainty about the quality, throw away the contents. Do not give the vegetables to animals.

Mixed vegetable ratatouille

A variety of vegetables cooked together and bottled makes a useful storecupboard item. Use juicy vegetables and cook without added water. Make up mixtures from whatever is available, but try to always have a base of tomatoes to give colour, flavour and moistness, and acidity, which is essential for safe keeping. Most of these dishes originated in the Mediterranean countries because of that area's profusion of vegetables.

The base:

1 kg (2 lb) ripe tomatoes
2 or 3 cloves garlic
juice of 1 lemon
salt, pepper
2 x 15 ml (1 rounded tablespoon) sugar

Add any 3 of the following:

2 aubergines
3 peppers (2 red, 1 green)
2 large onions
$\frac{1}{2}$ kg (1 lb) courgettes or marrow
1 cucumber
half a head of celery
2 cooked potatoes

Skin the tomatoes; hold over a gas flame or place in a bowl, cover with boiling water, leave for 1 minute and drain; then slice.

Peel the garlic, if used. Crush on a plate with a round-ended knife.

Prepare other vegetables. Slice the aubergines and sprinkle with salt to extract any water. Cut the peppers in halves, remove seeds, cores and pith, and slice. Peel and slice the onions. Wash and slice the courgettes, or cut the marrow into cubes. Cut the cucumber into thick slices, then in halves. Slice the celery.

In a large saucepan, fry the garlic and any onion. Add the lemon juice, sugar, salt and pepper and other vegetables; cover; and cook slowly, stirring occasionally, until the vegetables are tender, about 1 hour.

Pack into $\frac{1}{2}$-kg (1-lb) bottling jars; cover with lids and screw bands; and loosen these a quarter turn.

Place on the trivet of a pressure cooker; pour in warm water to a depth of 5 cm (2 in). Bring slowly up to 10-lb pressure and cook for 30 minutes. Leave to cool in the cooker for 10 minutes; then remove the bottles, tighten the screw bands, and place the bottles on a wooden surface to cool.

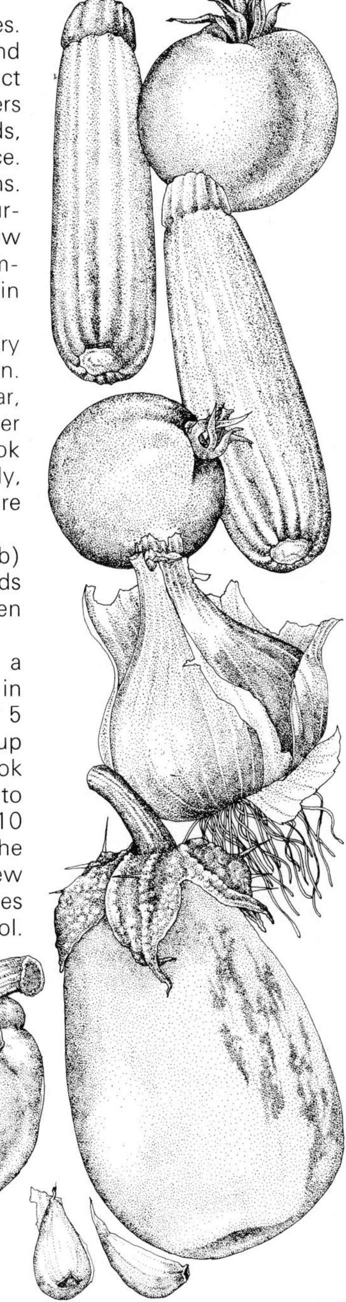

Pickling

A pickle is a fruit or vegetable preserved in spiced vinegar; sugar is sometimes added to give a sweet-and-sour flavour, but it is vinegar, which is added to prevent spoilage by micro-organisms, that is the all-important ingredient. Most gardens or allotments will provide enough surplus fruit and vegetables of the right kind for making pickles.

EQUIPMENT

Saucepan A large aluminium or stainless steel pan is best. Brass, iron or copper could be dangerous; also make sure that any enamel-ware used is unchipped.

Bowl A large earthenware bowl is required for brining the vegetables, or for mammoth quantities, a good-quality plastic washing-up bowl.

Grater Use a stainless steel grater for breaking down block salt.

Jars These must have plastic or plastic-coated tops to prevent the vinegar corroding the metal. Coffee jars with plastic tops are useful. Bottling or jam jars can also be used. Jam jars must be covered with snap-on plastic tops or tightly tied preserving skin to prevent evaporation of the vinegar. Fruit looks attractive packed in small jars fitted with corks.

INGREDIENTS

Firm varieties of fruits and vegetables are best for pickles, used while young and fresh. If they are to be pickled whole, choose small varieties for easy packing.

Vegetables Red or white cabbage is good for pickling; likewise cauliflower, cucumber, gherkins, onions, shallots, beetroot, walnuts, mushrooms; and mixtures.

Fruits Pickled fruits are usually spiced. Useful kinds include apples, apricots, bananas, blackberries, cherries, crab apples, damsons, gooseberries, grapes, lemons, peaches and pears.

Salt This is used to extract water so that the vinegar can penetrate and preserve the food. It also draws out starch from the cells of the food and makes it crisp. Block salt without any additives used to be best, but vacuum-packed ground salt is now also available. Avoid table salt which has substances added to prevent lumps, since they make the pickle cloudy. Iodized salt is also unacceptable as it gives an iodine flavour.

Vinegar Good-quality vinegar ensures a pickle that will mature well and will keep. There are many types available. To preserve food, the vinegar's contents must be at least 5 per cent acetic acid.

Malt vinegar This has a good flavour and is available both in the natural brown colour and as white distilled vinegar. The latter is useful for pale-coloured pickles; but the brown variety has the best flavour. Branded, bottled vinegar is preferable because it is usually of higher strength than bulk-packed vinegar. High-strength special spiced pickling vinegar has recently become available. This contains 8 per cent acetic acid, which speeds up the maturing time of the pickle and ensures that it keeps well. Bulk, draught or non-brewed vinegars, usually sold loose, should be avoided because they are generally of low strength.

Wine vinegar Inferior wine is used in making this, but the vinegar is usually of high strength.

Cider vinegar This is usually of high strength, and makes good pickles.

Spiced vinegar Spiced vinegar for pickling can be bought; but for individual flavour an original spice mix can be made. Use whole spices to avoid making the vinegar cloudy. Flavours take a long time to diffuse, which means that the spices must start being infused about two months before the vinegar is required. Drop the spices into a bottle of plain vinegar and leave them.

Mild spice mix For 1 litre (2 pints) of vinegar use a 5-cm (2-in) piece of cinnamon stick, 1 level tablespoon of whole cloves, 1 level tablespoon of black peppercorns, 1 level tablespoon of allspice berries (whole pimento or Jamaica pepper), and 1 level tablespoon of mace.

Hot spice mix Add 1 level tablespoon of chopped dried chillies and 2 level tablespoons of mustard seed to the mild spice mix.

BASIC METHOD FOR PICKLING VEGETABLES

1. Preparation of vegetables Choose young, fresh vegetables. Wash and drain them well, and remove any tough or damaged parts.

2. Brining For firm vegetables, soak in brine made by dissolving 50 g (2 oz) of salt in each $\frac{1}{2}$ litre (1 pint) of cold water for each $\frac{1}{2}$ kg (1 lb) of vegetables. Weight down the top with a plate to keep it immersed. Vegetables with a high water content such as marrow and cucumber should be layered in a bowl with dry salt. Leave the vegetables in the brine or salt for 24 hours.

3. Rinsing Drain the vegetables in a colander, and rinse under the cold tap; then leave them to drain thoroughly.

4. Packing Neatly pack the vegetables to the neck of each jar. Place with a wooden spoon handle if necessary.

5. Adding vinegar Top up with spiced vinegar. Use it cold for crisp vegetables and hot for soft ones. Cover the vegetables with vinegar to a depth of 1 cm ($\frac{1}{2}$ in).

6. Covering jars Make an airtight seal with plastic-coated screw tops, plastic tops, or preserving skin. A circle of thick plastic secured with a rubber band would also do. Jam-pot covers allow the vinegar to evaporate, and metal tops are corroded by vinegar. Label each jar, giving the date, variety of vegetable, and type of vinegar used.

7. Storage Store pickles in a cool, dark cupboard. Leave for one month for the flavours to blend if high-strength vinegar is used, and two to three months with ordinary pickling vinegar.

Recipes Many pickles can be made from a single variety of vegetable or fruit or from a mixture, either equally proportioned or with just a hint of a complementary flavour.

VEGETABLE PICKLES

Use metric or imperial quantities exclusively.

Beetroot

Wash, then boil in salted water (25 g or 1 oz of salt to $\frac{1}{2}$ litre or 1 pint of water) until tender, about $1\frac{1}{2}$ hours. Alternatively, cook at high (15-lb) pressure for about 30 minutes or bake in foil at 180°C, 350°F/Gas 4 for about $1\frac{1}{2}$ hours. Cook, skin,

then cut into rings or dice. Pack into jars and cover with cold spiced vinegar. Add 15 ml (1 level tablespoon) of salt to each $\frac{1}{2}$ litre (1 pint) of vinegar if not cooked in salted water.

Sweet beetroot pickle

Add 200 g (8 oz) of sugar to $\frac{1}{2}$ litre (1 pint) of vinegar.

Cabbage

Wash, shred, and layer with dry salt in a bowl. Leave for 24 hours; drain; rinse; drain thoroughly; then pack into jars. Cover with cold spiced vinegar, then seal. Use red cabbage after two weeks and hard white cabbage after one.

Cauliflower

Break into small florets, and leave in brine for 24 hours. Drain, rinse, drain once more and pack into jars. Cover

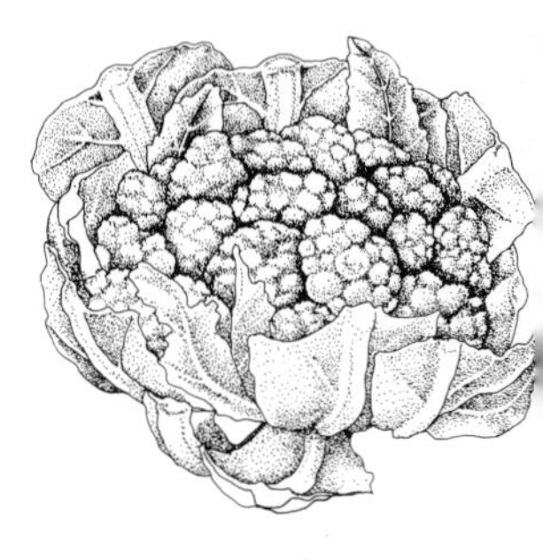

with cold spiced vinegar then seal.

Cucumber

If the cucumbers might be bitter, peel them. Cut into thick chunks. Layer with dry salt, leave 24 hours drain, rinse, drain again and pack into jars. Cover with cold spiced vinegar; seal.

Sweet cucumber pickle

Dissolve 200 g (8 oz) of sugar in each $\frac{1}{2}$ litre (1 pint of spiced vinegar over a low heat. Add the prepared cucumbers; boil; remove from the heat; and leave to cool. Pack the cucumbers into jars, cover with cold spiced vinegar, and seal.

Gherkins
Rub the gherkins with a cloth to remove the prickles. Wash, drain, and leave in brine for three days. Drain again and pack into jars. Cover with boiling spiced vinegar and leave in a warm place for 24 hours. Drain off the vinegar and boil. Repeat until the gherkins are a pleasant green colour. Add more spiced vinegar if necessary, and cover.

Mushrooms
Wash the mushrooms, and trim their stalks, but do not peel. If large, slice, or cut into quarters. Place in a casserole and cover with spiced vinegar. Add 10 ml (1 rounded teaspoon) each of salt and ground ginger and a shake of pepper for each $\frac{1}{2}$ kg (1 lb) of mushrooms. Cover, and cook in a slow oven (160°C, 325°F/ Gas 3) until the mushrooms are tender, about 45 minutes. Lift them out; drain; pack into jars; cover with hot vinegar before sealing.

Onions and shallots
Cover with brine; leave overnight. Drain; peel; cover in 2 litres (4 pints) of brine, for 24 hours if small; if larger, for 36 hours, placing a plate on top to hold them under the brine. Drain thoroughly. Boil the spiced vinegar in a large saucepan; add the onions or shallots; and boil for half a minute. Remove the onions; drain; and pack into warmed jars. Cover with hot vinegar Leave one month if using high-strength (8 per cent) pickling vinegar, or three months for (5 per cent) vinegar. Use distilled malt vinegar for silverskin onions.

Walnuts
Pick the walnuts before the shells have begun to harden. Test for this by pricking with a needle the part furthest from the stalk. Wear rubber gloves for this job because the nuts leave a brown stain. Cover the walnuts with standard-strength brine – 50 g (2 oz) of salt to $\frac{1}{2}$ litre (1 pint) of water for every $\frac{1}{2}$ kg (1 lb) of walnuts. Leave for three days; drain; and discard the brine. Cover the walnuts with more brine and leave for a week. Drain

thoroughly, spread on a large tray, and leave uncovered in a warm sunny room for about one day or until they have turned black. Pack into jars and cover with cold spiced vinegar. Store at least four weeks.

Sweet pickled walnuts
Prepare as above and cover with sweetened spiced vinegar made by boiling 25 g (1 oz) of bruised ginger, 25 g (1 oz) of white peppercorns, 25 g (1 oz) of whole cloves, and 25 g (1 oz) of allspice in $1\frac{1}{2}$ litres (3 pints) of malt vinegar. Cover and simmer for 15 minutes; then leave to cool. Strain off the spices, and stir $\frac{1}{2}$ kg (1 lb) of granulated sugar and two 15-ml spoons (1 rounded tablespoon) of black treacle or molasses into the vinegar.

MIXED PICKLES

Piccalilli
This can be mild or hot depending on the spices in its sauce. Its strength is determined by the proportions of mustard and ginger. Layer $1\frac{1}{2}$ kg (3 lb) of prepared vegetables with 200 g (8 oz) of salt. Combine cauliflower, pickling onions or shallots, marrow, courgettes, cucumber and French beans, cut into chunks. Place a plate on top and leave for 24 hours. Drain, rinse; and put in a saucepan with 1 litre ($1\frac{3}{4}$ pints) of distilled malt vinegar. Boil; cover; and simmer until the vegetables are just tender, about 20 minutes. Drain. Pack into warmed jars and cover with sauce.

Mild sweet sauce In a saucepan mix 15 ml (1 level tablespoon) of dry mustard, 5 ml (1 level teaspoon) of ground ginger, 15 ml (1 level tablespoon) of turmeric, and two 15-ml spoons (2 level tablespoons) of cornflour. Blend in 150 ml ($\frac{1}{4}$ pint) of vinegar, and add vinegar from the vegetables. Boil for 3 minutes, stirring continuously. Use the sauce to fill jars of vegetables.

Hot sharp sauce Make as above, but add three 15-ml spoons (3 level tablespoons) of mustard and 15 ml (1 level tablespoon) of ground ginger.

Mixed pickle
Use ridge cucumbers, cut into chunks, then halved; pickling onions or shallots; cauliflower florets; and French beans cut into pieces. Layer the vegetables with salt, and leave for 48 hours; then drain, rinse, drain again and pack into jars. Cover with cold spiced vinegar, and seal.

SWEET SPICY FRUIT PICKLES

These are a good accompaniment to roasted and cold meats, curries and cheese. The fruits are not brined, but preserved in sweet spiced vinegar. Wine, cider or distilled malt vinegar is used in preference to brown malt vinegar.

Choice of fruits for pickling
Firm or whole ripe fruits are best, such as apples, apricots, blackberries, cherries, black and red currants, damsons, gooseberries, grapes, peaches, pears, plums or rhubarb.

BASIC METHOD FOR PICKLING FRUITS

Preparing the vinegar

1. Dissolve 1 kg (2 lb) of granulated sugar in 1 litre ($1\frac{3}{4}$ pints) of wine, cider or distilled malt vinegar. Make up a mix of sweet whole spices; crush them lightly and tie them in a piece of muslin. Place in the sweetened vinegar; bring to the boil; and leave to infuse while preparing the fruit.

2. Preparation of fruits Peel the apples and pears, stone the cherries, and top and tail the gooseberries. Hard fruits like damsons, if they are not to be cut, need pricking with a stainless steel or silver fork for the flavour to penetrate.

3. Place the fruit in a saucepan with the sweet vinegar. Bring to the boil, cover, and cook very gently until the fruit is just tender but not broken up. Drain the vinegar syrup and pack the fruit into jars.

4. Boil the syrup in an uncovered pan until slightly thicker; then pour into the jars to cover the fruit by at least 1 cm ($\frac{1}{2}$ in). Cover and tie down. The fruit absorbs the syrup, so keep any surplus syrup in a covered jar for topping up. Store the fruit for two months.

Sweet spice mix
To spice $\frac{1}{2}$ litre (1 pint) of wine, cider or distilled malt

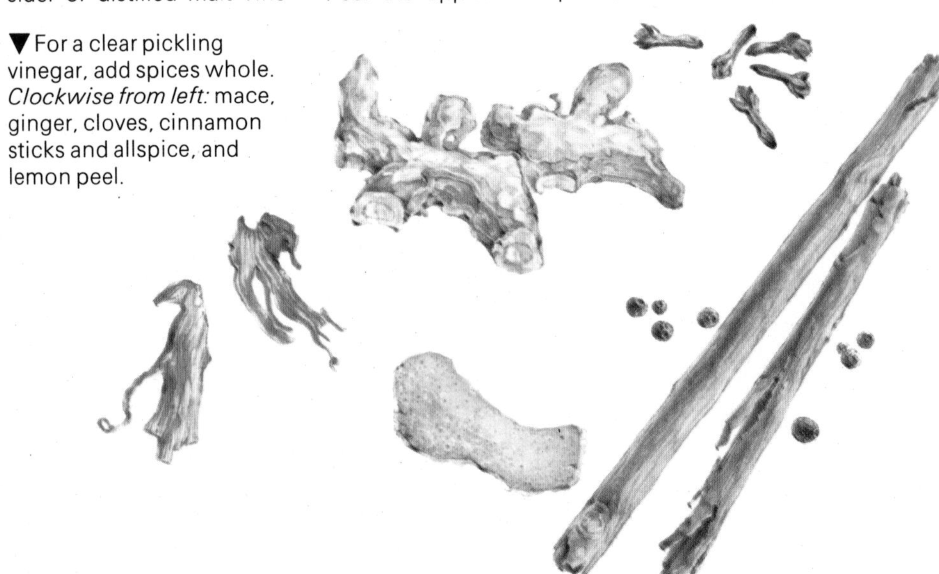

▼For a clear pickling vinegar, add spices whole. *Clockwise from left:* mace, ginger, cloves, cinnamon sticks and allspice, and lemon peel.

vinegar, sufficient for 1 to 2 kg (2 to 4 lb) of fruit, add:

1 kg (2 lb) sugar
2 blades mace
15 ml (1 level tablespoon) allspice berries
15 ml (1 level tablespoon) coriander seeds
15 ml (1 level tablespoon) whole cloves
½ cinnamon stick
strip of lemon rind (optional)
piece of root ginger (optional)

Spiced apple pickle
Use firm, acid apples. Prick the fruit or cut them in half; cook in sweetened, spiced vinegar until just tender; and pack into jars. Reduce the vinegar by boiling and pour it over the apples. Seal the jars.

Spiced apricot pickle
Cut the apricots in halves and remove the stones. Place the fruit, cut side downwards, in a large, shallow saucepan; cover with sweet spiced vinegar; cook slowly until just tender; then pack into jars. Reduce the vinegar by boiling for about 5 minutes; pour it over the fruit; cover; and seal.

Spiced damson pickle
Prick the damsons with a stainless steel or silver fork. Cook in spiced, sweet vinegar until just tender. Drain off the vinegar and pack the damsons into jars. Reduce the vinegar by boiling for 5 minutes; then pour it over the fruit. Leave overnight; then drain off the vinegar again. Repeat four times on successive days; then seal.

Spiced pear pickle and **spiced peach pickle**
As for apricot pickle.

Pickled eggs
Around Easter, when eggs are cheap, preserve them this way.

Use eggs about two days old. Hard-boil them for 10 minutes, giving the pan an occasional stir to keep the yolks in the centre. Plunge them into cold water; then, when cool, remove the shells. Pack loosely into large glass or stone jars and cover with cold spiced vinegar. Shake the jar gently now and then, because if the eggs are stuck together they will develop a mottled appearance. The colour of the vinegar is important to the appearance of pickled eggs; so do some with brown malt vinegar, some with white wine vinegar, and some with red wine vinegar to achieve a pink colour. To spice wine vinegar, boil each ½ litre (1 pint) with a rounded tablespoon of whole pickling spice for 5 minutes; then leave to cool before straining.

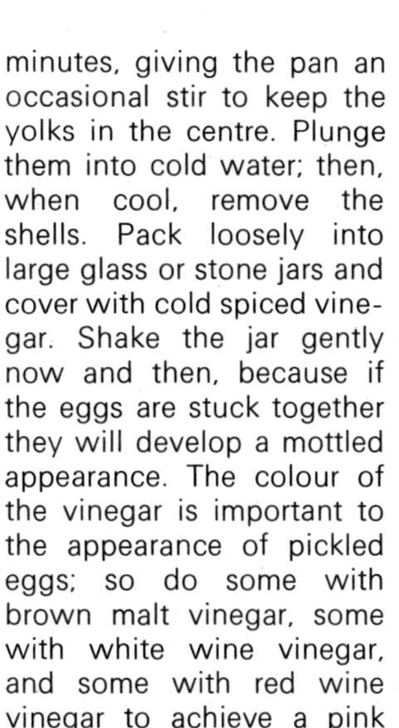

▲ Pack pickled hard-boiled eggs in spiced vinegar. Use different kinds of vinegar to vary the colour of the eggs.

What went wrong?
Pickle shrunk in jar
Cover not airtight. Paper covers or jam covers are not suitable.
Cloudy vinegar
Vegetable not left long enough in the brine. Spices not strained sufficiently.
Yellow spots on pickled onions
Harmless to eat, but it looks unsightly. Blanch the onions for half a minute in boiling vinegar and pack while hot.

Chutneys

Chutneys are made from fruits, vegetables or a mixture of both. They contain both sugar and vinegar to preserve them and to give a sweet-and-sour flavour. They can include spices that are either aromatic and mild or hot and pungent. The recipes given are only approximate guides, leaving the cook scope to experiment with different spice mixes. It is as well, however, to make only a small quantity until the flavour is perfected. Spices mellow with age; so chutney benefits from being left for a couple of months before being eaten.

EQUIPMENT

Very little special equipment is necessary.

Saucepan The saucepan is the most important item. Stainless steel, aluminium or enamel pans are best. It is essential that the enamel pan is not chipped. Never use containers of copper, brass or iron. A well-fitting lid is useful for chutneys that need long, gentle cooking without too much evaporation.

Mincer Ingredients need to be broken down so that they blend and cook to a pulp. A mincer is especially useful for this.

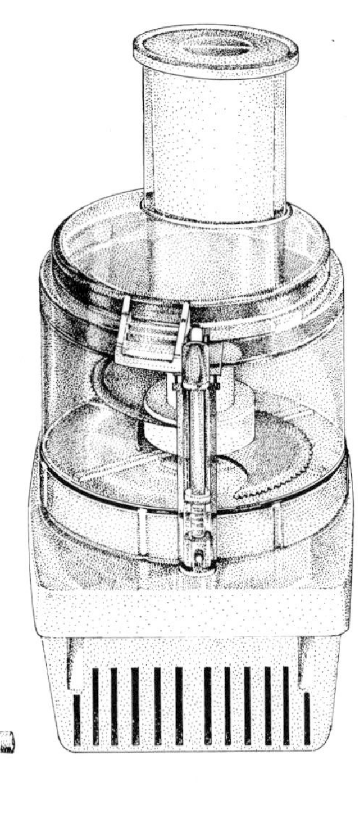

Chopping board and knife These provide for the simplest method of preparing the ingredients, but also the most laborious.

Electric chopper The luxury gadget that makes short work of preparation; useful for hard ingredients.

Sieve Use a nylon sieve or a stainless steel strainer, but not a metal sieve.

Funnel The same funnel can be used as for jam. The jars can however be filled, very carefully, without one.

Jars When a jar is opened, evaporation takes place, so choose small jars and use up the chutney fairly quickly.

Covers As with pickles, the vinegar corrodes metal and for this reason any metal tops must be plastic-lined. Use plastic screw or snap-on tops, double-thick plastic tightly held with a rubber band, or preserving skin.

Labels To give details of the chutney, make your own labels from sticky paper or punched tape, or buy labels from specialist shops or stationery departments.

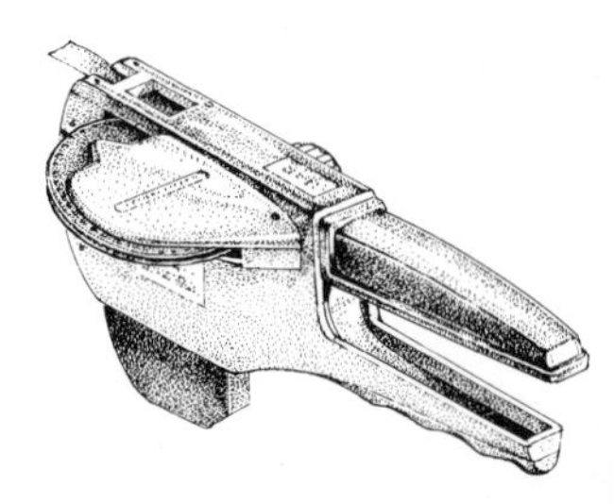

INGREDIENTS

Fruits and vegetables Because chutney ingredients are usually reduced to a pulp, they can conveniently include any blemished or misshapen fruits or vegetables. Use up windfall apples, green tomatoes or end-of-season coarse sticks of rhubarb. The most useful ingredients are apples, onions, plums, red and green tomatoes, marrow, garlic and shallots.

Ingredients for flavour Ingredients added to give flavour include spices, salt, dried fruits, vinegar and sugar. Spices are usually added ground up. It is important to make sure they are fresh because after grinding they quickly deteriorate. If using whole spices, tie them in muslin and cook them with the fruits and vegetables. Malt vinegar gives the best flavour; but use distilled malt vinegar for pale-coloured chutneys of mild flavour. Either brown sugar or white can be used. Black treacle and molasses give chutneys a rich flavour and dark colour, but they should be added in moderation.

BASIC METHOD FOR MAKING CHUTNEY

1. Preparing the ingredients Chop or mince the vegetables and fruits.

2. Cooking Aim to cook very slowly to extract the flavour and make a mellow chutney. It may be necessary to cook tough ingredients first in a covered saucepan or pressure cooker and then add the remaining ingredients and cook slowly for at least 2 hours with the lid off. For a light-coloured chutney, add sugar, and some of the vinegar, near the end of the cooking time. Boil the mixture until thick, with no free liquid.

3. Potting Ladle into clean, warmed jars. Cover with an airtight, vinegar-proof top. Wipe and label the jar, giving the variety of chutney and the date.

▲ Use surplus marrows, apples, onions and tomatoes for making chutney.

CHUTNEY RECIPES

These can be varied to suit individual tastes. The varieties of spices or dried fruits can be changed, but keep the quantity of fruit, vegetable, vinegar and sugar the same in order to ensure a good keeping chutney. For a hot chutney use ground ginger, mustard and curry powder, about 15 ml (1 level tablespoon) of each, and 5 ml (1 level teaspoon) of cayenne to each 2 kg (4 lb) of mixed fruit and vegetables. Reduce these quantities for a mild chutney. Use cinnamon, coriander, allspice, nutmeg, mace and cloves for sweet, aromatic chutneys. Leave the finished result to mellow for at least two months before eating.

Use metric or imperial quantities exclusively.

Apple chutney

Apples make a good base for chutney, giving a smooth, thick texture.

2 kg (4 lb) apples
1 kg (2 lb) onions
4 cloves garlic
2 x 15 ml (1 rounded tablespoon) salt
½ kg (1 lb) sultanas
2 x 15 ml (1 rounded tablespoon) ground cinnamon
2 x 15 ml (1 rounded tablespoon) mustard
5 ml (1 level teaspoon) cayenne
1 kg (2 lb) sugar
1 litre (1½ pints) vinegar

Peel and core the apples; add the peeled onion and garlic; then mince or chop. Place in a saucepan with the salt, sultanas, cinnamon, mustard, cayenne and half the vinegar and cook slowly, stirring occasionally, until soft and pulpy, about 1½ hours. Add the sugar and remaining vinegar and cook until thick. Pot and cover while hot.

This makes about 3 kg (6 lb) of chutney.

Sharp apple chutney

2 kg (4 lb) apples
½ kg (1 lb) onions
3 oranges
3 lemons
2 x 15 ml (1 rounded tablespoon) curry powder
2 x 10 ml (1 rounded dessertspoon) cloves
200 g (8 oz) seedless raisins
1 litre (1½ pints) vinegar
½ kg (1 lb) granulated sugar
4 x 15 ml (2 rounded tablespoons) black treacle

Peel, core and chop the apples; and peel and chop the onions. Scrub the oranges and lemons, and pare the rinds thinly, avoiding any pith. Chop the peel and squeeze out the juice, and place in a saucepan with the curry powder, crushed cloves, raisins and half the vinegar. Cook until pulpy, stirring occasionally, for about 1 hour. Add the sugar, black treacle and remaining vinegar; bring to the boil; and cook until thick. Pot and cover while hot.

This makes a quantity of about 3 kg (6 lb).

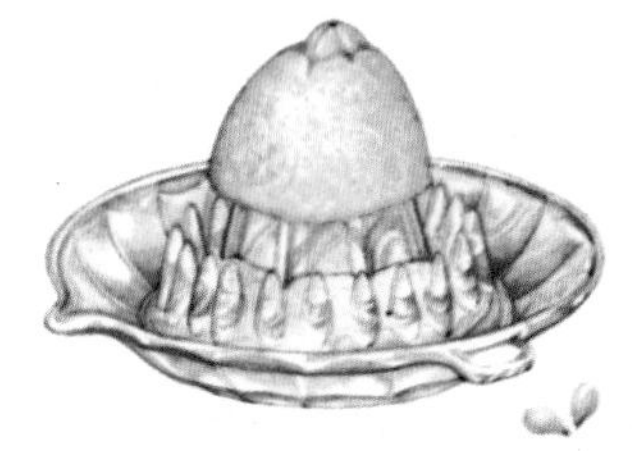

Blackberry and apple chutney

3 kg (6 lb) blackberries
1 kg (2 lb) cooking apples
½ kg (1 lb) onions
15 ml (1 level tablespoon) ground nutmeg
15 ml (1 level tablespoon) ground cinnamon
5 ml (1 level teaspoon) cayenne

10 ml (1 rounded teaspoon) salt
1 litre (1½ pints) vinegar
1 kg (2 lb) sugar

Place the blackberries in a saucepan. Wash the apples, and chop without peeling; put them in the pan with 250 ml (½ pint) of water and cook slowly until the fruit is pulped. Rub through a sieve, rinse the pan, and return the purée to the pan. Peel and chop the onions, and add to the purée with the spices, the salt and half the vinegar. Cook until the onion is tender; then add the sugar and remaining vinegar. Bring to the boil, stirring frequently. Cook slowly until thick. Pot and cover while hot.

▶ Tomato and courgette chutney has a sweet, spicy and fruity flavour, perfect with bread and cheese.

Chopping an onion

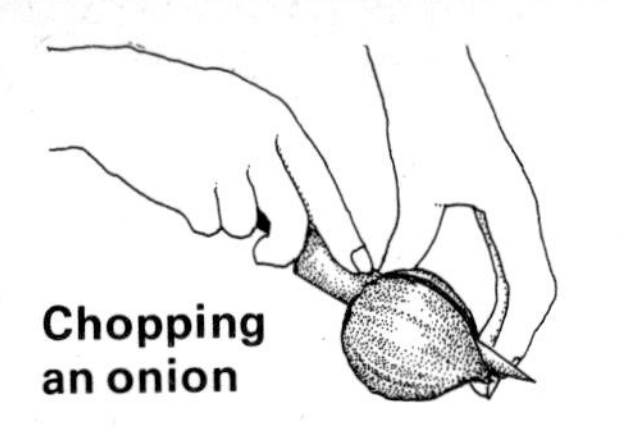

1. Cut the onion in half.

2. Cut each half in strips.

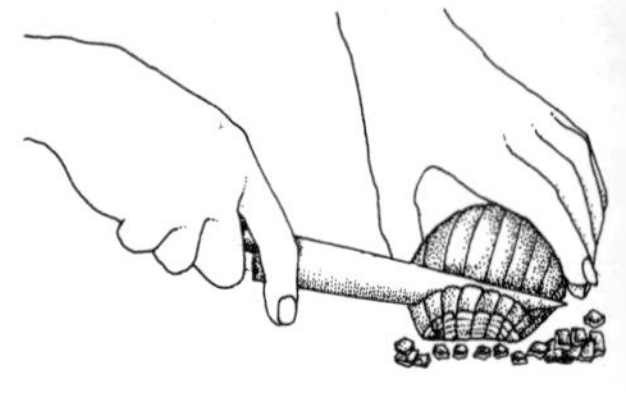

3. Chop the onion into cubes

Brown spicy chutney

1 large orange
1 kg (2 lb) green or red tomatoes
$\frac{1}{2}$ kg (1 lb) apples
$\frac{1}{2}$ kg (1 lb) carrots
1 kg (2 lb) onions
25 g (1 oz) garlic
250 g (8 oz) stoned dates
50 g (2 oz) salt
15 ml (1 level tablespoon) ground cloves
5 ml (1 level teaspoon) cayenne
15 ml (1 level tablespoon) coriander
1 litre (2 pints) vinegar
1 kg (2 lb) sugar
100 g (4 oz) black treacle

Squeeze the juice from the orange; cut up the peel. Peel red tomatoes, apples, carrots, onions and garlic; chop or mince with the orange peel and dates. Cook in a covered saucepan with the salt, orange juice, spices and half the vinegar, stirring occasionally, about 2 hours. Add the remaining vinegar, sugar and treacle. Bring to the boil, stirring; cook until thick and pulpy. Pot and cover while hot.

Pear chutney

1 kg (2 lb) prepared pears
$\frac{1}{2}$ kg (1 lb) onions
$\frac{1}{2}$ kg (1 lb) red or green tomatoes
250 g (8 oz) sour apples
250 g (8 oz) celery
100 g (4 oz) seedless raisins
2 x 15 ml (1 rounded tablespoon) pickling spice
1 litre (1$\frac{1}{2}$ pints) vinegar
$\frac{1}{2}$ kg (1 lb) sugar

Quarter the pears, then halve the pieces crosswise. Peel and chop the onions; peel any red tomatoes; and quarter the tomatoes. Peel, core and chop the apples; wash and chop the celery; and chop the raisins. Securely tie the pickling spice in a piece of muslin. Place these ingredients in a large saucepan with half the vinegar and boil until tender, about 2 hours, stirring occasionally. Very hard pears can be cooked in a covered pan, or in a pressure cooker at high (15-lb) pressure. Add the remaining vinegar and sugar; boil until thick; and remove the spices. Pot and cover.

Rhubarb chutney

2 kg (4 lb) rhubarb
$\frac{1}{2}$ kg (1 lb) onions
250 g (8 oz) stoned dates
1 kg (2 lb) sugar
15 ml (1 level tablespoon) salt
15 ml (1 level tablespoon) ground ginger
15 ml (1 level tablespoon) curry powder
1 litre (1$\frac{1}{2}$ pints) vinegar

Prepare the rhubarb: peel the onions; and finely chop or mince them both. Chop or mince the dates. Place all ingredients in a saucepan and cook slowly until thick. Pot and cover while hot.

Tomato and courgette chutney

3 kg (6 lb) ripe tomatoes
1 kg (2 lb) courgettes
$\frac{1}{2}$ kg (1 lb) onions
15 ml (1 level tablespoon) paprika
15 ml (1 level tablespoon) ground cinnamon
5 ml (1 level teaspoon) mixed spice
2 x 15 ml (2 level tablespoons) salt
400 ml ($\frac{3}{4}$ pint) distilled malt vinegar
$\frac{1}{2}$ kg (1 lb) sugar

Skin the tomatoes, cut into quarters. Wash the courgettes and cut into chunks. Peel and chop the onions. Cook slowly with the spices and salt, until pulpy, about 2 hours. Add the sugar and vinegar and cook until thick. Pot and cover while hot.

What went wrong?

Chutney shrunken in jar
Cover not airtight, or of the wrong material to prevent evaporation.

Liquid on top of chutney
Insufficient boiling to evaporate liquid.

Sauces and ketchups

Today's smooth sauces developed from the spiced vinegar left after a pickle had been finished; frugal cooks preferred not to throw these flavourings away and they were termed "store sauces". Old cookery books advise that the ones most likely to be needed are mushroom and walnut ketchups, essence of anchovies, chilli, cucumber, shallot, and Harvey or Worcester sauce, both now available commercially. A ketchup is thinner than a sauce and is the savoury juice from a fruit or vegetable. It is highly seasoned, with one predominating flavour.

EQUIPMENT

This is the same as for cooking chutneys (see p. 62), except that a nylon or hair sieve is required. As with chutneys, avoid the use of metal. A sieve attachment on an electric mixer is a good help if sauces are to be made in quantity. An electric chopper or mincer is useful too, because the ingredients need to be finely divided for the flavour to be extracted.

Bottles and jars For smooth sauces like tomato, bottles are needed. Small 250-ml (8-oz) non-returnable bottles from mixer drinks are ideal, but they must be capable of being sterilized. Save old sauce bottles with screw tops; and, for concentrated mint and horseradish sauces, keep any ready-mix mustard jars.

Large saucepan Tomato and mushroom ketchups must be sterilized. A pan is needed deep enough for the bottles to be placed upright in it and covered with water.

Grater A stainless steel grater is necessary for making horseradish sauce.

INGREDIENTS

Many fruits are good basic ingredients: tomato is the favourite; but apples, blackberries, damsons, plums, red currants and elderberries also make delicious sauces. Choose ripe, fully-flavoured fruits. Other types of sauces are the concentrated ones in which mint and horseradish are preserved in vinegar. Mushrooms and walnuts make excellent ketchups. Vinegar, onions, garlic and spices are used for added flavour.

BASIC METHOD FOR SMOOTH SAUCES AND KETCHUPS

1. Prepare the fruit or vegetable, but leaving the peel, skin, cores or stones, to be removed, if preferred, during sieving or straining. Cut or chop the fruit finely.

2. Cook the fruit or vegetable in its own juice or with some of the measured vinegar until soft. Rub through a nylon or hair sieve for sauces; and strain for ketchups.

3. Add the spices, vinegar and sugar and boil until, for sauces, the mixture has the consistency of thick cream.

4. Pour into hot bottles; seal with screw tops or corks; then sterilize in a saucepan of water. Place the bottles on a wad of newspaper; fill the pan with hot water; slowly bring up to simmering point; then keep at this temperature for 30 minutes. Remove from the pan; tighten the tops; then cool and label.

5. Store in a cool, dark cupboard.

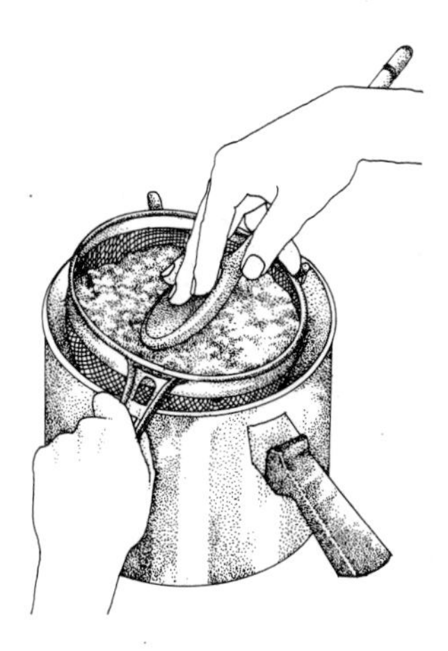

▶ Use a nylon sieve for acid fruits and vegetables. Press the bowl of a wooden spoon with the fingers to help push the ingredients through.

RECIPES FOR SAUCES AND KETCHUPS

As with the chutneys, these recipes are only a guide. Mix and match with whatever is available. For hot sauces, use chilli, ginger, cayenne and mustard; for gentler flavours, use cinnamon, cloves, mace and allspice. Use spiced vinegar for convenience.

Use metric or imperial quantities exclusively.

Ripe tomato sauce

This sauce is only lightly spiced, so that the fresh, ripe flavour of the tomatoes is not masked. Since it uses a lot of tomatoes, if they are not available replace 1 kg (2 lb) of their weight with apples.

5 kg (12 lb) ripe red tomatoes
2 sticks celery or 15 ml (1 level tablespoon) celery salt
50 g ($1\frac{1}{2}$ oz) salt
$\frac{1}{2}$ kg (1 lb) sugar
2·5 ml ($\frac{1}{2}$ level teaspoon) cayenne pepper
15 ml (1 level tablespoon) paprika
$\frac{1}{2}$ litre (1 pint) spiced vinegar

Wash the tomatoes and celery; chop them; and place in a large saucepan. Cook slowly until pulped, then press through a sieve. Rinse the pan, place the purée in it and add the other ingredients. Bring to the boil, stirring frequently, and cook until the mixture has thickened and has the consistency of cream. Pour into bottles, and fit screw tops or tie-down corks. Place in a saucepan of water; sterilize for 30 minutes (see p. 67).

Tomato ketchup

3 kg (6 lb) ripe red tomatoes
2 sticks celery or 15 ml (1 level tablespoon) celery salt
1 large onion
2 large cloves garlic
2·5 ml ($\frac{1}{2}$ level teaspoon) cayenne pepper
5 ml (1 level teaspoon) ground ginger
15 ml (1 level tablespoon) salt
5 ml (1 level teaspoon) ground cinnamon
15 ml (1 level tablespoon) ground coriander
150 ml ($\frac{1}{4}$ pint) distilled vinegar
250 g (8 oz) sugar

How to peel tomatoes

Pour boiling water over the tomatoes; count to ten; then drain them.

For a large quantity, heat water in a saucepan. See that at some point each fruit is immersed.

The skins should peel off easily. Hold each tomato steady with a fork.

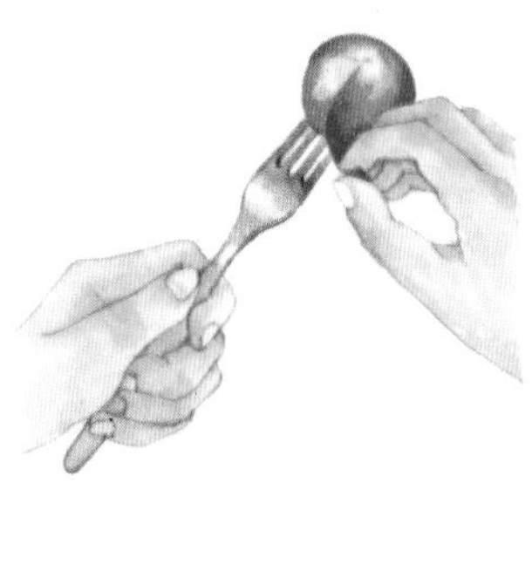

To peel a few tomatoes, spear each one on a fork and scorch the skin over a gas flame.

Wash the tomatoes and any celery. Peel the onion and garlic. Chop the ingredients. Cook slowly until pulped; then sieve. Rinse the pan, and boil the purée and remaining ingredients in it, meanwhile stirring, until they have the consistency of thin cream. Bottle while hot, and sterilize.

Concentrated horseradish sauce
For really pungent horseradish, dig up a root. It is often to be found growing wild. Grating horseradish is even worse on the eyes than peeling onions, so do it near an open window, or outside if you can. Place ½ litre (1 pint) of water in a saucepan, add 5 ml (1 level teaspoon) of salt, and bring to the boil. Wash and scrape a horseradish root; grate finely and immediately place in the salted water to preserve its colour. Alternatively, chop with water in a blender. Drain, and pack into warmed jars. Cover with boiling distilled malt vinegar and seal.

▼ Learn to recognize horseradish.

▲ Store grated horseradish in vinegar in small jars ready for use.

To make horseradish sauce
Drain the horseradish in a concentrated sauce from the vinegar and mix with thin cream or thick, mild mayonnaise and sugar to taste.

Mushroom ketchup

Large field mushrooms have the best flavour for ketchup, but must be gathered on a dry day for it to keep.

1 kg (2 lb) mushrooms
50 g (2 oz) salt
5 ml (1 level teaspoon) black peppercorns
5 ml (1 level teaspoon) allspice berries
10 cloves
piece of root ginger
$\frac{1}{2}$ litre (1 pint) malt vinegar

Break up the mushrooms, place them in a bowl and mix in the salt. Cover, and leave overnight. Mash with a potato masher or wooden spoon, and place in a saucepan. Crush the spices with the end of a rolling pin; add, with the vinegar, to the pan; bring to the boil; cover; and simmer for 30 minutes. Strain through a nylon sieve into a jug; then bottle and seal while hot. Sterilize the bottles, and label when cool.

Elderberry ketchup

Most old cookery books have a recipe for this sauce. It is sometimes called "Pontac's sauce" after the owner of a famous eating house in the City of London. Originally it was made with claret, but when prices rose, frugal cooks substituted vinegar.

1 kg (2 lb) elderberries
6 shallots
$\frac{1}{2}$ litre (1 pint) vinegar
5 ml (1 level teaspoon) salt
15 ml (1 level tablespoon) black peppercorns
15 ml (1 level tablespoon) allspice or cloves
2·5 ml ($\frac{1}{2}$ level teaspoon) cayenne

Strip the elderberries from their stalks, and peel and slice the shallots. Place together in a casserole; boil the vinegar with the spices; add it to the casserole and cook in a slow oven (150°C, 300°F/Gas 2) until the elderberries are tender, for about 3 hours. Leave to cool, and strain through a sieve, but do not press. Pour into bottles, seal and sterilize.

Plum relish

Dark red plums are the most attractive in this thick sauce. Serve it with turkey, duck, or cured meats such as bacon.

$1\frac{1}{2}$ kg (3 lb) plums
200 g (8 oz) onions
200 g (8 oz) seedless raisins
1 medium-sized orange
2 × 15 ml (1 rounded tablespoon) salt
15 ml (1 level tablespoon) ground coriander
10 ml (2 level teaspoons) mustard
2·5 ml ($\frac{1}{2}$ level teaspoon) cayenne pepper or chilli powder
10 ml (1 rounded teaspoon) cloves
$\frac{1}{2}$ litre ($\frac{3}{4}$ pint) vinegar
$\frac{1}{2}$ kg (1 lb) sugar

Wash the plums, cut in halves, and if possible remove the stones. Peel and chop the onions, and chop the raisins. Scrub the orange; squeeze out its juice; cut out some of the white pith; then chop the peel. Place together in a saucepan with the spices and half the vinegar. Cook until pulpy, for about 30 minutes. Rub through a sieve, return to the saucepan and add the remaining vinegar and sugar. Bring to the boil, stirring. Cook for about 10 minutes until thickened; then fill the bottles; seal and sterilize.

Concentrated mint sauce

Pick fresh mint leaves on a warm day just before the plant flowers. Chop finely and pack loosely into small jars. Cover with cold distilled malt vinegar, or wine or cider vinegar. Seal, and store in a cool, dark place. To use, place 2 teaspoons of mint in a bowl, and add 1 teaspoon of sugar and a little boiling water to dissolve the sugar. Add more vinegar, if desired.

▼ Several varieties of mint can be used for sauce. *Left:* spearmint. *Right:* peppermint.

Freezing food

Preserving food by cold, as in short-term refrigeration or long-term freezing, is the most natural way of keeping it, for unlike other preservation methods it does not alter the structure of the food. Its disadvantage is the initial high cost of the equipment needed to maintain sufficiently low temperatures.

How cold?
We cannot assess the degrees of cold by feel in the same way as degrees of heat, though it is possible to burn both food and one's skin by excess cold as well as by excess heat. The degree of cold is depicted on the equipment by the star marking system (see below). Commercial models freeze food very quickly at very low temperatures: −34°C, −30°F. Storage time depends directly on the temperature. For example, vegetables suitably packed will keep for three days in the main body of the refrigerator; in its frozen food compartment for a week, a month or three months at one-, two- or three-star temperatures respectively; and for up to a year in a freezer.

EQUIPMENT
Choosing a freezer The storage space usually dictates the type of freezer. The following are available.
1. Upright There are one or two front-opening doors to this type; one compartment can be a refrigerator. The freezing coil can be hidden round the side or built into a shelf. Choose a model with the coil in the shelf if

Frozen food compartment of refrigerator

*	−6°C, 21°F
**	−12°C, 10°F
***	−18°C, 0°F

Main body of refrigerator
4°C, 40°F

Freezer

−18°C, 0°F for storage
−21° to −24°C, −5° to 12°F for freezing fresh food

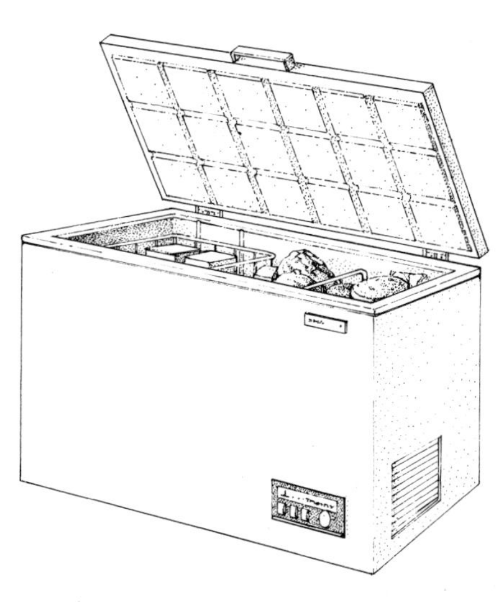

you plan to freeze your own food. Upright freezers are easy to stack food in and keep tidy.

2. Chest This is the familiar white coffin with a top-opening lid. It is ideal for the storage of large packages. It is marginally cheaper to buy and to run than an upright freezer, but requires more organization to keep it tidy and manage the stock of food. Short people find it difficult to reach the bottom of a chest freezer. When you buy a freezer, check that it is capable of freezing fresh food. It should have the freezer symbol, and the temperature should go down to −24°C, −12°F. Avoid buying a conservator; because since its temperature will not go below −18°C, 0°F, it can only store previously frozen food rather than preserve fresh food.

Size Food must be frozen as quickly as possible. The ice crystals formed should be small, to hold the juices in the cells of the food. Slow freezing makes big ice crystals, which rupture the cells and cause the moisture to run out of the food as it thaws. A powerful freezer is necessary for freezing fresh food in quantity. A home freezer can only freeze one tenth of its capacity each day – that is, 1 kg for each 28·3 litres ($2\frac{1}{2}$ lb per cubic ft). Large freezers are usually best value, since extra running costs are negligible. For storing pre-packed and prepared foods, 350 litres (12·4 cubic ft) is the smallest convenient size, because the packs are large.

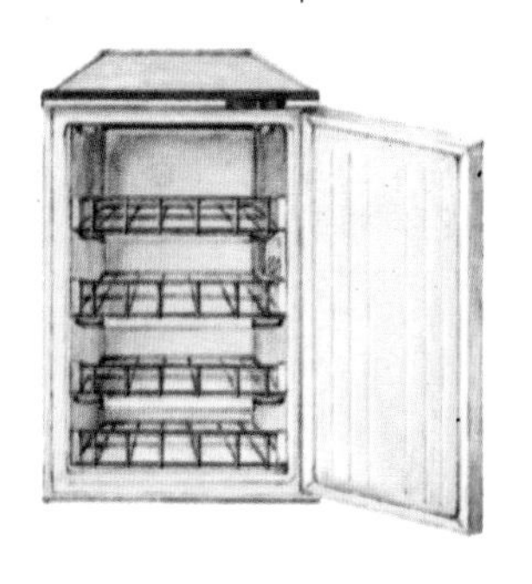

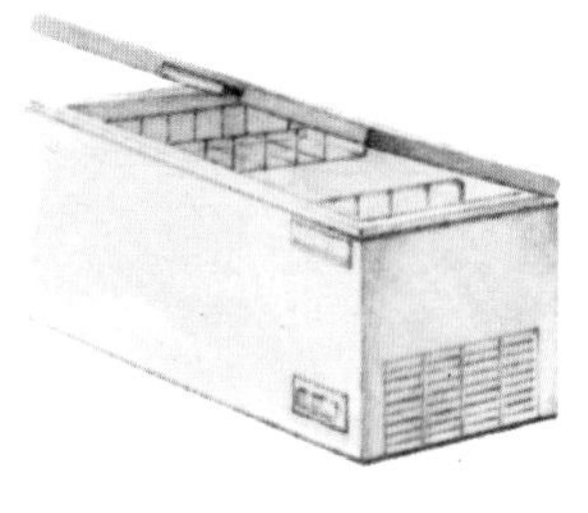

Packaging materials

Since the cold air of a freezer has a very drying effect, food needs to be protected with moisture-proof packaging. For long-term storage, this must be quite thick. Most freezer packaging catalogues contain a comprehensive list of suitable car-

Pack food for freezing in moisture-proof materials. Oblong containers make best use of the space.

tons, bags, and rolls of wrapping materials. It can be useful to save plastic cartons, foil plates and trays and thick polythene bags from commercial frozen foods. Re-usable plastic boxes with well-fitting lids are initially expensive, but they are very easy to use and last for years. Foil containers can also be re-used. Use a waxy chinagraph pencil to write on them. For other packages use tie-on or stick-on labels.

BASIC METHOD FOR FREEZING FRUIT

If you are planning to freeze more than ½ kg (1 lb) of food, operate the fast-freeze switch. Choose firm, ripe fruits. It is not worthwhile to freeze any but the most perfect, unless intended for jam, chutney or wine. There are three methods of freezing fruit:

1. Loose or open freezing

Use this method for free-flow packs. Choose raw fruits that do not discolour. Those with hard skins, like black and red currants, plums, damsons and gooseberries, can be packed dry in freezer bags. Wrap oranges for marmalade individually in self-clinging plastic; then pack in a freezer bag. Fruit for jam can be packed in a freezer bag as it is. Label it, giving weight and intended use. Freezing destroys some of the pectin, so pack an extra 100 g (4 oz) of fruit for each 1 kg (2 lb). Soft fruits must be spread in a single layer on metal trays. Chill in the refrigerator; then leave in the coldest part of the freezer – on the base or a shelf with a freezer coil. To avoid drying, aim to freeze the fruit in about 1 hour. Pack immediately.

2. In dry sugar

Use 200 to 300 g (8 to 12 oz) of castor sugar to each 1 kg (2 lb) of fruit. Either lightly turn the fruit in the sugar to coat it, or layer it in a plastic container.

3. In syrup Make a syrup, using 250 g to ½ kg (8 oz to 1 lb) for each 550 ml (1 pint) of water; and chill. Use about 250 ml (½ pint) for each ½ kg (1 lb) of fruit. Leave 2 cm (1 in) of the container unfilled to allow for expansion during freezing. Cover, label and chill; then freeze. Use this method for fruits that need cooking, or ones that discolour, like apples, apricots and peaches. As an extra precaution against discoloration, add some lemon juice or Vitamin C to the syrup. To keep the fruit under the syrup and prevent the top layer discolouring, place a ball of greaseproof paper on top. If containers are limited, remove the frozen block, wrap in self-clinging plastic, and store in a freezer bag. Alternatively, line a cardboard box with a freezer bag; fill; freeze; then remove the box.

Layer the fruit with sugar or cover with syrup. Place crumpled greaseproof paper on top to hold the fruit under the syrup.

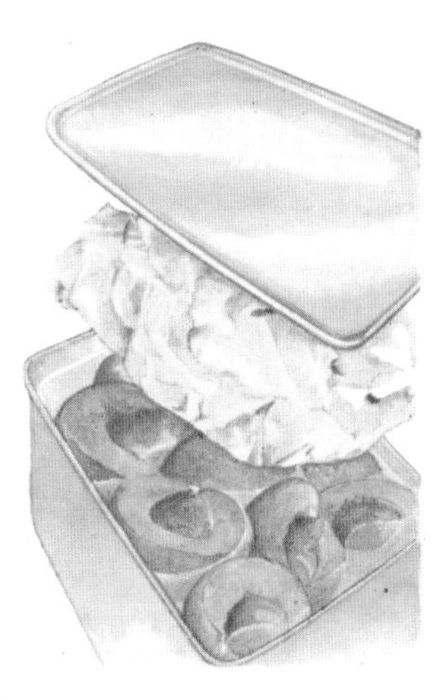

FREEZING FRUIT

Fruit	Preparation	Freezing method	To use
apple	**1.** Peel, core, slice into salted water. Place in muslin bag or perforated container; blanch 2 minutes; drain	In dry sugar	Stew from frozen Thaw for pies
	2. Make a purée	In containers	Thaw
apricots	Remove stones; poach in syrup to prevent browning	In syrup	Thaw
bilberries black currants red currants	Wash only if necessary	Dry in bags	Stew from frozen Thaw for pies
cherries	Wash only if necessary; remove stones	In syrup	Poach if necessary Thaw
gooseberries	Top and tail	Dry in bags or in syrup	Stew from frozen Thaw for pies
grapefruit and oranges	In segments Juice	In dry sugar In tubs or ice trays	Thaw Thaw
melon	Peel; remove seeds; cube	In syrup	Thaw
peaches	Dip in boiling water; remove skins and stones; cut in halves or slices	In syrup with lemon juice or Vitamin C	Thaw
pears	Not worth freezing. During a glut, peel, core, and cut into slices	In syrup with added lemon juice or Vitamin C	Poach
pineapple	Peel, core, and cut into slices or cubes	In syrup	Thaw
plums	Avoid washing, if possible. Stew if very ripe	Dry in bags In containers	Stew from frozen Thaw for pies Thaw
soft fruits (blackberries loganberries raspberries)	Avoid washing if possible	Open-freeze on trays; then pack in bags or boxes	Thaw very slowly in refrigerator
strawberries	Only suitable for cooking, as they collapse. Best as purée	Best in syrup As purée, in containers or ice trays	Thaw very slowly in refrigerator
tomatoes	Leave whole, to be used for cooking since they collapse. Cook and make into purée	Whole, dry in bags As purée, in containers	Grill or fry from frozen Thaw or heat

BASIC METHOD FOR FREEZING VEGETABLES

Freeze only young, fresh vegetables. A possible exception is a soup or stew-pack of raw or cooked vegetables to avoid waste. Chill until ready to freeze. Operate the fast-freeze switch for over ½ kg (1 lb) of food.

1. Preparation Prepare according to kind, and cut into even-sized pieces.

2. Blanching To inactivate enzymes spoiling colour and flavour, the vegetables must be blanched in boiling water. Only a small quantity can be done at a time: the boiling water must regain its temperature quickly. Omit blanching only if storing less than two weeks.

Prepare a large saucepan about three-quarters full of boiling water. Place ½ kg (1 lb) or less of the vegetables in a wire blanching-basket. Improvise, if necessary, with a nylon bag (as used for straining wine) or the separators from your pressure cooker. Lower the vegetables into the water and time from where it returns to the boil (see chart). Cool in ice-cold water for the same time; then spread on clean tea towels and pat dry. The blanching water can be used about five times.

3. Freezing Open-freeze vegetables on trays, or pack in plastic boxes or freezer bags in amounts suitable for the most usual servings. Press out any air with the hands, or use a pump; close with a wire tie; label. Place on a freezing shelf or in the fast-freeze compartment.

(a) French beans being cut to a convenient length.

(b) Blanching the beans in boiling water.

(c) Cooling the blanched vegetables in iced water.

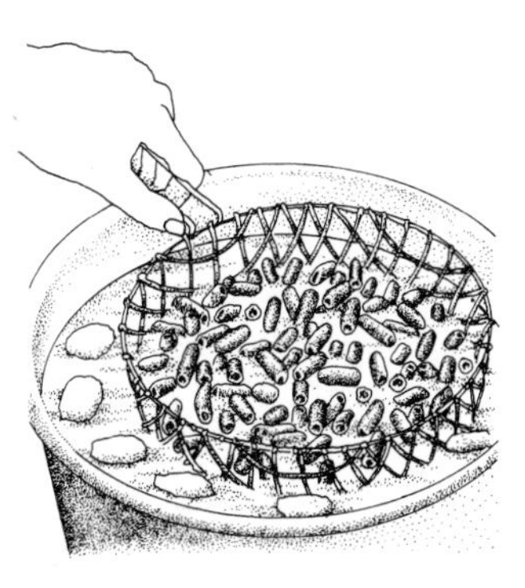

(d) After chilling in the water, the vegetables are patted dry.

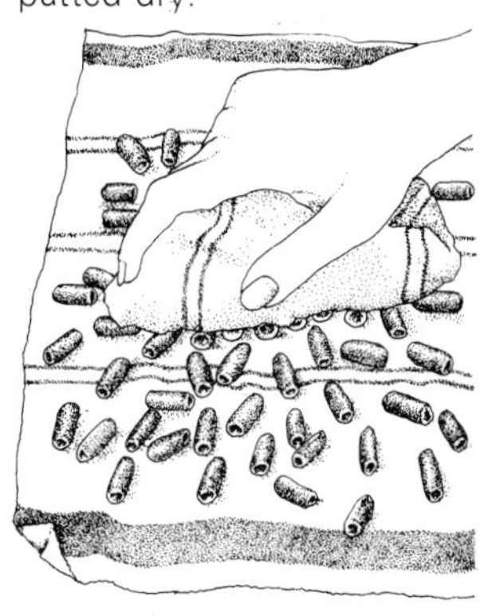

(e) Pack in boxes in usable amounts.

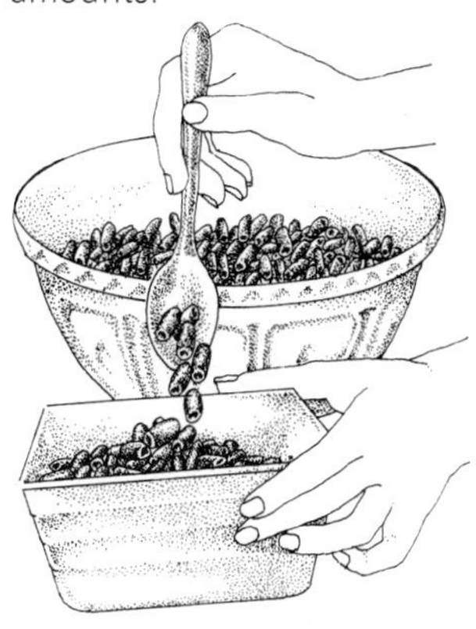

Processed sprigs of broccoli being packed ready for the freezer.

Vegetable	Preparation	Blanching and cooling time	Freezing method
asparagus	Grade for size	thin 2 minutes thick 4 minutes	Pack into containers
beans:			
broad	Shell, and grade for size	small 2 minutes large 3 minutes	Pack in freezer bags
French	Trim ends; cut in chunks if desired	whole 3 minutes cut 2 minutes	Open-freeze; pack in bags
runner	String, if necessary; cut into chunks	2 minutes	As French beans
broccoli	Grade for size; soak 30 minutes in cold salt water	3 to 4 minutes	Open-freeze; pack in boxes
Brussels sprouts	Grade for size: small sprouts are best. Soak 30 minutes in cold salt water	3 minutes	Open-freeze; pack in bags or boxes
carrots	Freeze only small, young carrots; scrub and trim	5 minutes	Rub off skins; pack in bags
cauliflower	Break into florets	3 minutes	Open-freeze; pack in bags
courgettes	Cut into 2-cm (1-in) lengths	3 minutes	Open-freeze; pack in bags
mushrooms	Wash and dry; cut in quarters if large	(**a**) Fry in butter (**b**) Do not blanch	(**a**) Pack in containers (**b**) Open-freeze
peas	Shell and grade into sizes	1 minute	Open-freeze
peppers	Wash; cut in halves; remove seeds and membrane; slice	2 minutes	Pack into bags or boxes
potatoes:			
new	Choose small potatoes	Until almost cooked	Pack in bags
chipped	Fry until almost cooked but not browned	Cool quickly	Open-freeze; pack in bags
spinach	Prepare as for serving	2 minutes; stir	Pack in bags
mixed vegetables	Prepare each vegetable separately; then mix	Blanch separately	Pack in bags

▶ Fragile fruits need packing in containers; and those that brown easily must be covered with syrup. Protect the protruding bones of poultry with foil. Heavy-gauge polythene bags are economical for packing.

BASIC METHOD FOR FREEZING MEAT AND FISH

The most convenient frozen packs of meat and fish are those which are either in sufficient portions for the meals served or those where single portions can be removed. It should not be necessary to saw frozen blocks of food; it is no good for the food or the temper. Remember to freeze only 1 kg (2 lb) for each 28·3 litres (1 cubic ft) of freezer space in any 24 hours, or the quality of the meat will suffer.

Wrapping chops, steaks and fillets Interleave with plastic tissue or greaseproof paper. Wrap the whole package in foil and press out any air. If necessary, seal with freezer tape. Label, giving name, number of portions and date of freezing; chill; and freeze. Alternatively, pack each chop, steak or fillet in self-clinging plastic wrap and place in a freezer bag. Small, whole gutted fish are best packed this way for protection.

Cook from frozen.

Wrapping joints Aim to make the pack as airtight as possible. Cover in self-clinging freezer wrap, and stick on a label or place in a labelled freezer bag. Thaw before cooking.

Wrapping poultry and game Rinse and dry thoroughly. Make a pad of crumpled foil or greaseproof paper to cover any protruding bones. Wrap as for joints. Thaw thoroughly before cooking in the usual way.

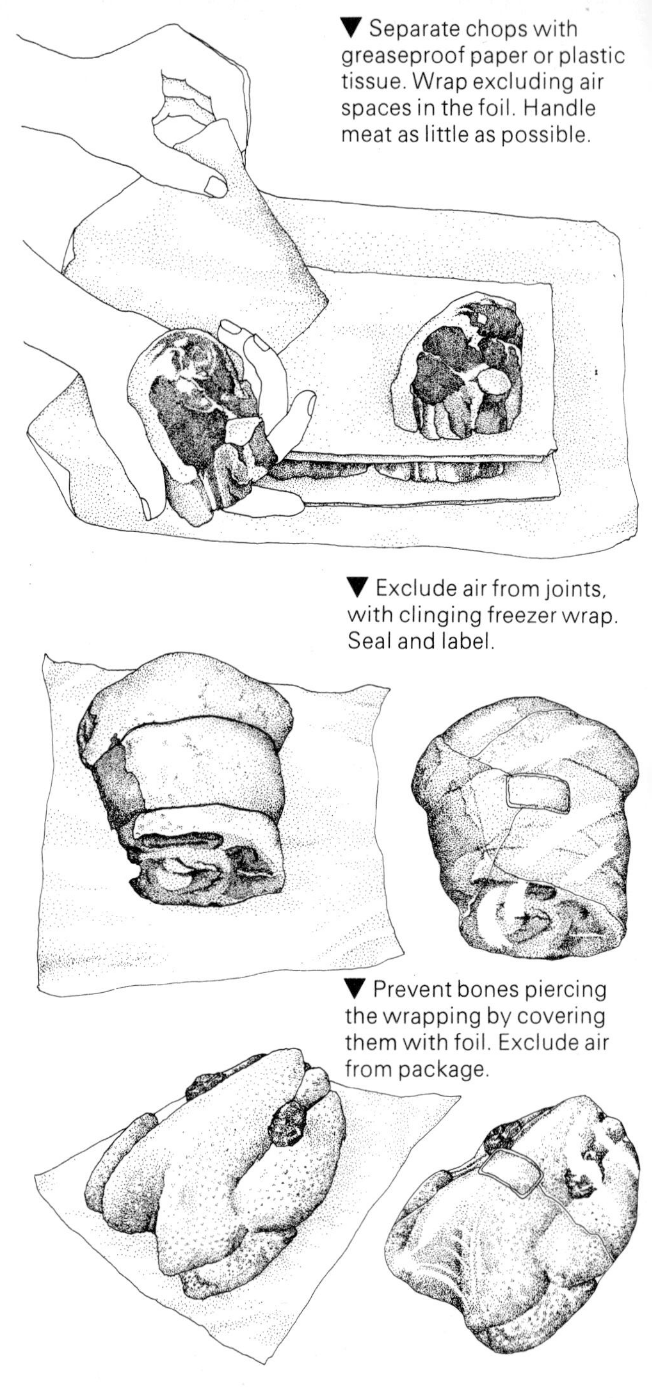

▼ Separate chops with greaseproof paper or plastic tissue. Wrap excluding air spaces in the foil. Handle meat as little as possible.

▼ Exclude air from joints, with clinging freezer wrap. Seal and label.

▼ Prevent bones piercing the wrapping by covering them with foil. Exclude air from package.

▼ Small, whole fish store best when glazed. Open-freeze scaled, gutted fish; then dip each one in iced, salted water. Return it to the freezer; then repeat until it is covered with ice. Pack the fish individually in cling wrap.

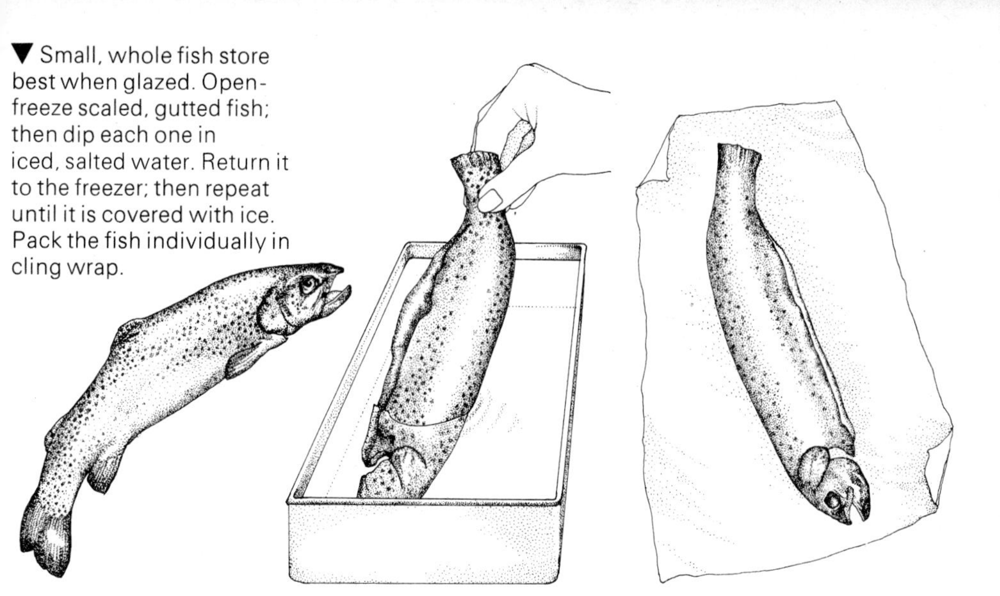

THAWING FROZEN FOODS

Small pieces of meat and vegetables are best cooked from frozen. Food that needs thawing before cooking, such as poultry or joints of meat, is best thawed very slowly. This way, the moisture is re-absorbed by the tissues of the food. If time is short, meat can be cooked from frozen but it will lose a lot of moisture and become dry. Poultry must always be thawed, so that the temperature of the inside is raised sufficiently to kill any harmful bacteria. For this reason, stuff the neck, never the body. If you have run out of time, joint the bird after thoroughly cooking.

If a freezer breakdown has partially thawed the food, it can be safely re-frozen if still icy. Flavour will have deteriorated slightly.

STORAGE TIME OF FROZEN FOODS

Air and fluctuating temperatures are the main causes of deterioration in frozen foods. Fatty foods cannot be kept for long because they start going rancid. Cured foods such as bacon, unless vacuum-packed, have a short storage time.

To attain the maximum storage time of frozen foods:

1. Keep in airtight packs and exclude as much air as possible.

2. Freeze as quickly as possible; make sure the package being frozen is in contact with the freezing coil and is not touching any other frozen packages. As well as slowing down the freezing, this would also make the already frozen food deteriorate.

3. Switch to 'fast freeze' if freezing more than 1 kg (2 lb) of food.

4. Do not overload your freezer — check the manufacturer's instructions.

As a general rule, fruits and vegetables store for 1 year; beef and lamb 1 year; pork 9 months; white fish 6 months; oily fish (salmon, trout, mackerel, herrings) 4 months.

What went wrong?

Dry, discoloured patches
Packaging torn or not moisture-proof.

Frosted food in packages
Temperature fluctuation, or air not expelled.

Dry meat or fish
Freezing process too slow.

Flavour deterioration
Fluctuating temperature.

Dry storage of fruit and vegetables

The simplest method of storing some apples and pears, and some vegetables too, is just in air. No expensive equipment is required for this means of food preservation, but it is necessary to make sure that the variety chosen is known to be a keeper.

STORING FRUIT

Most late season's pears and apples, if perfect, will store well. Good dessert apples include Cox's Orange Pippin, Blenheim Orange, Laxton's Superb, and Sturmer Pippin; the best cookers are Bramley's Seedling and Newton Wonder. Pears that store well are: Glou Morceau, Josephine de Malines and Winter Nelis.

Choice of fruit The apples and pears must be fully ripe. Pick the apples carefully to avoid bruising. Select only those that are perfect; preserve by other methods or use immediately any that have peck marks or scab or any other defect. Place the apples in a single layer and leave in a cool, airy place overnight.

Packing Wrap each fruit separately in waxed paper, tissue or even newspaper to delay the ripening process and prevent decay. Place in a single layer in boxes, pears with the stalk uppermost, apples with the stalk down. Cover each layer to keep out the light. Boxes can be stacked; but there must be room for air to circulate.

Where to store A garage, garden shed or outhouse is ideal. If it is likely to be frosty, protect the fruit with layers of straw or newspaper. In mild climates, the boxes can be stored outside; but they must be well protected from frost and covered with wire netting to keep out mice. The storage place must not have a warm or drying atmosphere or the fruit will shrivel.

STORING VEGETABLES

Root vegetables Beetroots, carrots, turnips, parsnips, swedes and potatoes will all store well throughout the winter.

Preparation Harvest vegetables before the first frosts. If they are left too long, the roots might split. Sort carefully and store only undamaged roots. Twist off the tops and layer the vegetables, a little apart, with dry sand or peat in boxes or in a plastic dustbin. Store this in somewhere cool and

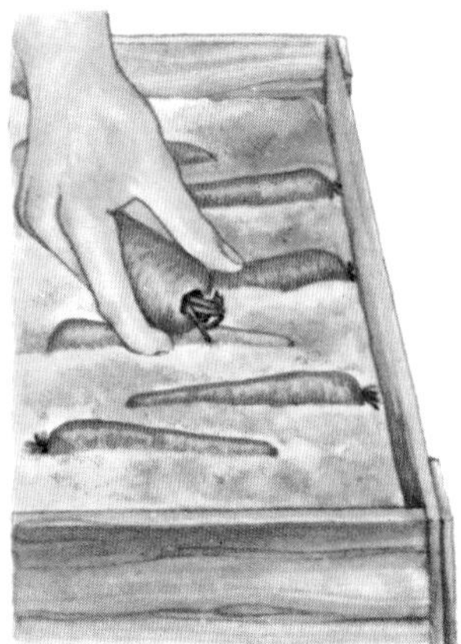

frostproof, and avoiding a very dry atmosphere. Potatoes must be kept dark. If they are in boxes, cover with black polythene.

Marrows and squashes For storage, pick large mature marrows and squashes. Hang them in string bags in a frostproof place.

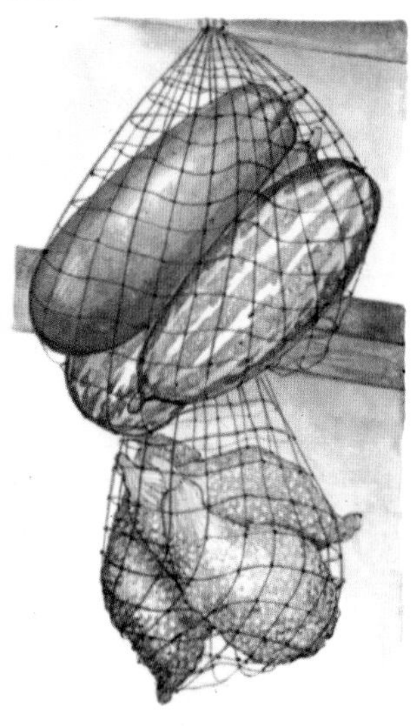

Onions, shallots and garlic Store when ripe and dry – shown by withering of leaves at the plant's base. Hang up shallots in string bags. Store garlic in the same way, or, as for onions, plaited round a rope, then hung in a cool, dry place away from frost. Check the vegetables periodically, and remove any soft ones.

STORING NUTS

Store only mature nuts.

Chestnuts Make sure they are sweet chestnuts and not horse chestnuts (the leaf of the sweet chestnut tree is pointed and single). Remove the husks; and wipe the nuts and store in boxes or sacks in a cool, dry place.

Walnuts Collect immediately they drop, and clean as soon as possible. Remove the husk and any fibre on the shell by scrubbing quickly. Avoid the nut becoming too wet. Spread out the nuts in a single layer and leave to dry in a warm, airy room. Pack in a large crock, separating the nuts with a mixture of salt, to prevent mould, and coconut fibre, to absorb moisture. Keep in a cool, dark place and remove as required.

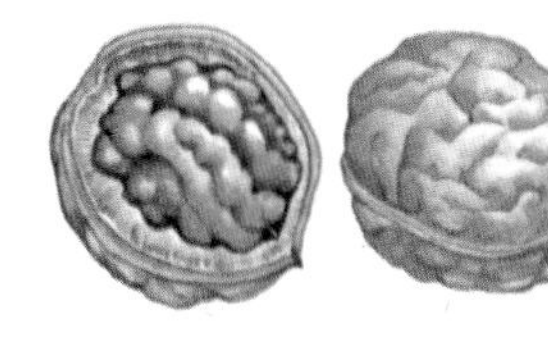

Cobnuts Spread out on trays to dry, turning occasionally.

Drying fruit, vegetables and herbs

The oldest and most straightforward means of preserving food entails drying it, by exposure to heat and freely circulating air. It is still practised today, mainly by more sophisticated methods than were available to our ancestors. In many processes aimed at drying food for storage the use of an oven has replaced the unreliable heat of the sun.

EQUIPMENT

Trays Food to be dried needs to be in a single layer and to have a free circulation of air. A fine-mesh tray such as a wire cooling rack is necessary, though if the pieces of food are small, the mesh needs to be very fine. If food is being dried in quantity, it is probably worthwhile to make a wooden frame and stretch a fine fabric (such as cheesecloth or a piece of sheeting) across. Make sure that the fabric is easily removable, because it will need boiling before and after use. Check that the frame can fit into the oven and allow sufficient air circulation around it.

detachable fabric base

wooden frame

Oven A gas, electric or solid-fuel cooker is adequate, but it must be capable of being switched to a very low heat, 70°C, or 150°F/ Gas $\frac{1}{4}$. Use the simmering oven of a solid-fuel cooker.

FRUITS FOR DRYING

Apples, pears and plums dry best and are also the easiest fruits to obtain. Choose unblemished ripe and juicy fruits for easy drying and the best flavour.

BASIC METHOD FOR DRYING FRUITS

1. Prepare the fruit; cut apples into rings, pears into quarters, and halve plums and remove their stones. Place apples and pears in a bowl of salt water, using 50 g (2 oz) of salt to $4\frac{1}{2}$ litres (1 gallon) of water, to prevent browning.

2. Dry the apples and pears on kitchen paper or a clean cloth, and spread the fruit on a mesh tray. Apple rings can be threaded on a bamboo pole.

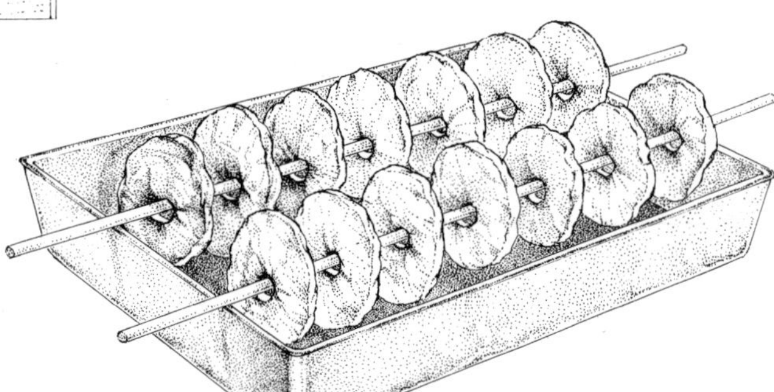

▼ Apple rings prepared for drying.

3. Dry the fruit in an oven, setting its heat as low as possible and leaving the door open to give a good draught of air. The temperature of the oven must not exceed 70°C, 150°F/ Gas $\frac{1}{4}$, or the flavour of the fruit will spoil. Leave for about 6 hours. The heat need not be continuous.

4. Test the fruit by pressing between the fingers. There should be no moisture apparent.

5. Leave for 12 hours, then pack in layers between waxed paper in a ventilated cardboard box. Store in a cool, dry place.

To use dried fruit

Soak the fruit overnight in a bowl of water. Bring slowly to the boil in the water used for soaking; cover; and simmer until tender. Stir in sugar, if necessary.

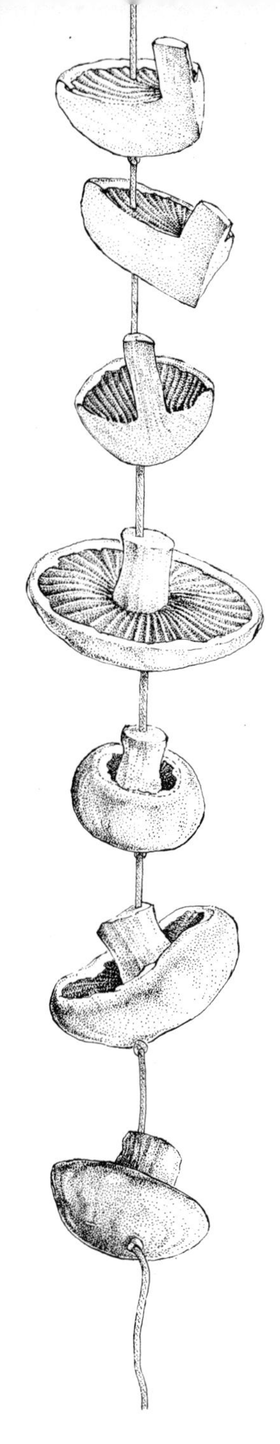

VEGETABLES FOR DRYING

Most vegetables can be dried successfully; but as they cannot re-hydrate back to their original succulence, other methods of preservation are preferable. Runner beans and mushrooms can be dried with most success. Peas, and mature and starchy varieties of beans such as haricot and soya, are best left on the vines to dry slowly.

BASIC METHOD FOR DRYING VEGETABLES

1. Prepare the vegetables: runner beans, for example, need to be finely sliced.

2. Blanch runner beans by placing in boiling water for 2 minutes; drain; dry on kitchen paper or a clean cloth; then spread on a mesh tray.

Mushrooms should be washed only if necessary. Place on a mesh tray, thread on fine string, or spike on to a twig. Dry in a cool oven, not set at more than 70°C, 150°F/Gas $\frac{1}{4}$, until crisp. Dry the twig with mushrooms in the airing cupboard or in the sun. Leave to cool; then pack in tightly corked jars.

To use dried vegetables

Soak overnight; then add salt and boil until tender. Mushrooms can be crumbled into casserole dishes.

◀ Mushrooms can be separated for drying by threading on a string or by spearing on a twig.

DRYING HERBS

Pick the herbs when they are full of flavour, just before they flower. Choose a warm, dry day, before the sun is too hot to evaporate the essential oils that give the flavour.

Tie small-leafed herbs such as thyme, parsley and winter savoury in bundles, and hang them up to dry in a warm airy place. A garden shed or well-aired cupboard would be ideal. If it is dusty, cover the herbs with a loosely woven cloth.

Strip larger-leafed herbs from their stalks and place on a mesh tray. Dry in a very cool oven on the lowest setting, not more than 70°C, 150°F/Gas $\frac{1}{4}$. Leave to cool; then crumble them. To mix a herb blend with a unique flavour, add a very little quantity of dried lavender to a blend of thyme, sage, marjoram and basil.

Grated horseradish can also be dried. Spread on mesh trays until it crumbles when pressed with the fingers.

Storage Store dried herbs in dark bottles with screw tops. Replace every year.

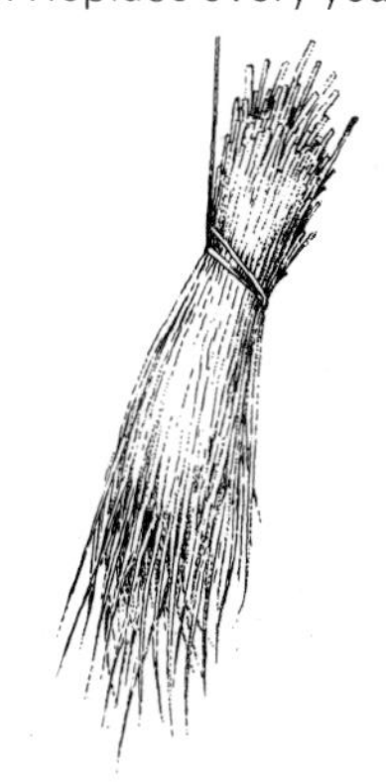

Preserving with alcohol

Preserved fruits, as well as many fresh ones, can be used with spirits and sugar to make a great variety of drinks and special desserts. Liqueurs are just infusions of flavours in spirits, and involve no fermentation. For fruit desserts, brandy is popular; but the spirits that can be used are almost unlimited.

DESSERTS

Choose full-flavoured, acid fruits, to balance the flavour of sugar and alcohol.

Using spirits

Brandy has one of the flavours most compatible with fruits; but many spirits can be used. Full-flavoured fruits like morello cherries, raspberries, loganberries and pineapple are delicious with kirsch or white rum. Use dark rum with full-flavoured mixed fruits. In France, *eau-de-vie*, made specially to go with fruit, is sold at a reasonable price in the supermarkets.

Strong expensive spirits are not necessary. Instead of five-star brandy, the low-alcohol grape variety of about 40 per cent proof can readily be used.

Brandied cherries

Use large, ripe morello cherries. If using canned or bottled fruit, overlook the cooking directions.

½ kg (1 lb) morello cherries
250 g (8 oz) sugar
150 ml (¼ pint) brandy

Trim the stalks of the cherries to 1 cm (½ in). Wash the cherries; then prick all over with a large needle or cocktail stick. Dissolve half the sugar in 300 ml (½ pint) of boiling water. Add a few cherries at a time, bring to the boil and simmer for 1 to 2 minutes. Take care that the skins do not burst. Remove the cherries with a draining spoon; pack into small, wide-necked jars, repeat. Add the remaining sugar and stir until dissolved. Boil until the syrup forms beads on the surface. It should register 216°F or form a thread when pressed between two teaspoons which are then pulled apart. Measure the syrup; cool; then add an equal amount of brandy. Pour over the cherries, making sure they are well covered. Seal with a screw or clip top. Store about four months.

Brandied peaches

Both fresh and canned fruit can be preserved this way; also apricots, pineapple and plums. Choose a sharp, firm peach such as Hale.

½ kg (1 lb) peaches
250 g (8 oz) sugar
150 ml (¼ pint) brandy

Halve and stone the peaches. In a large shallow pan, make a syrup with half the sugar and 300 ml (½ pint) of water. Poach the peaches until just tender, turning once. Remove from the pan with a draining spoon and skin them. Pack the fruit neatly into wide-necked jars. Dissolve the remaining 125 g (4 oz) of sugar in the syrup. Boil to the thread stage and finish the fruit as for brandied cherries.

Rumtopf

German housewives make this preserve as the fruit becomes available, to store until Advent. Traditionally, attractive blue-patterned stone jars are used.

½ kg (1 lb) each of sour cherries, strawberries, apricots, peaches, plums, pears, pineapple
castor sugar
dark rum

Prepare the fruit: prick the skins of the cherries, apricots and plums; halve the apricots; slice the peaches; skin and slice the pears; remove the skin from the pineapple, cut out its core, and cut the pineapple into chunks. As each variety becomes available, sprinkle with 100 g (4 oz) of castor sugar to each ½ kg (1 lb) of fruit and leave overnight. With strawberries use 200 g (8 oz) of sugar. Place in a rumtopf jar or a large glass jar, and cover with rum, then with a lid. Repeat with each fruit, finishing with a layer of pineapple. Cover, and leave for three months.

LIQUEURS

Liqueurs are easier to make than wines. Instead of being fermented, they are made by infusing particular flavours in spirits.

INGREDIENTS

Fruit Use strongly flavoured fruits such as raspberries, loganberries, black currants, blackberries, apricots, peaches, damsons, oranges and lemons.

Spirits The cheaper grape brandies or lower-alcohol "cocktail" gins are adequate. *Eau-de-vie* can also be used.

Spices Add spices whole, to avoid clouding the liquor, after lightly bruising them.

Sugar White sugar is included in most liqueurs. If its colour is an advantage, demerara sugar can be used.

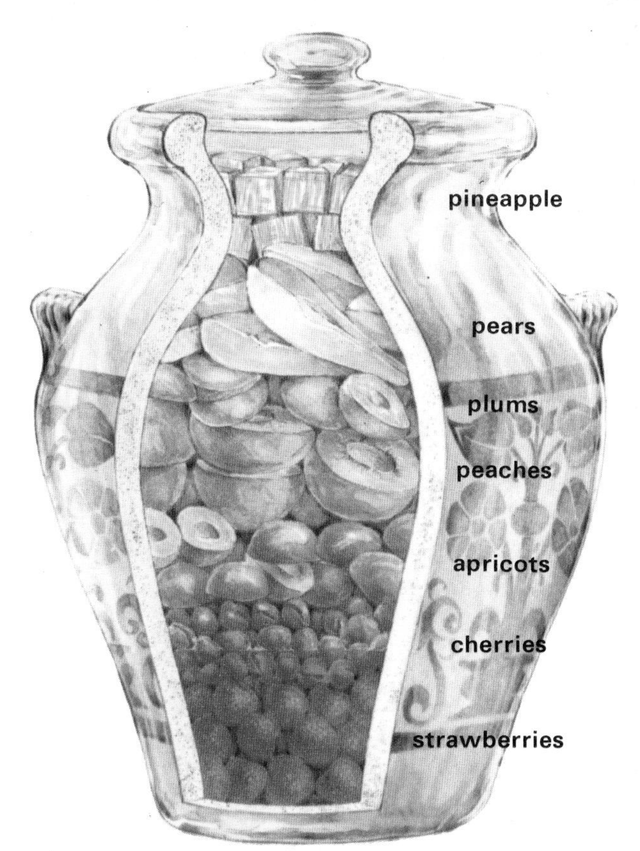

Raspberry gin

Use this basic method for strawberry, loganberry and black-currant gin too.

½ kg (1 lb) raspberries
350 g (12 oz) sugar
¾ litre (26⅔ fl oz) bottle of gin

Place the ingredients in a jar; cover with a lid, or a double thickness of plastic sheeting; and tie this down. Leave for three months, shaking every day for the first month, then occasionally. Strain; pass through a filter paper into bottles; then fit the corks.

Sloe gin

Sloes are the fruit of the wild blackthorn. They need pricking to extract the flavour. Make up a pad of spikes for this from wooden or plastic cocktail sticks. Place the pricked sloes in a jar; then proceed as for raspberry gin.

Peach brandy

½ kg (1 lb) ripe peaches
250 g (8 oz) demerara sugar
½ litre (1 pint) brandy

Wash the peaches and slice finely. Place in a jar, and add the sugar and brandy. Crack 3 peach stones; chop the kernels, and add them. Cover with a lid, or a double thickness of plastic sheeting, and tie down. Leave for one month. Shake the jar daily for two weeks; then occasionally. Strain, pass through a filter paper into bottles; then fit corks.

Orange whisky

Use this method for making flavoured whiskies from other citrus fruits. They can be mixed; include Seville oranges to give a tang.

2 medium-sized oranges
150 g (6 oz) granulated sugar
½ litre (1 pint) whisky

Scrub the oranges, and pare the rind very finely, avoiding any pith. Squeeze out the juice. Place in a jar with the other ingredients; cover with a lid or with a double thickness of plastic sheeting; and tie down. Store for one month, shaking the jar every day for two weeks then occasionally. Strain; pass through filter paper into a bottle; and fit a cork.

Metrication

Both metric and imperial quantities are given in the recipes and methods in this book. The exact conversion of 1 oz to 28·349 grams makes an impossible calculation, and to make it easier, 1 oz has been rounded to 25 grams (abbreviated as g), for small quantities. For large quantities, 1 lb has been rounded to ½ kilogram (abbreviated as kg). All the amounts have been rounded in this way and the recipes have been tested using both amounts. As they are equivalent proportions and not equivalent amounts, it is important to use *either* the metric quantities or the imperial quantities. Please do not mix them. The chart of equivalents is as follows when large amounts are being used:

1 kg (1,000 g) = 2 lb
½ kg (500 g) = 1 lb
¼ kg (250 g) = ½ lb (8 oz)
125 g = 4 oz
50 g = 2 oz
25 g = 1 oz

For small amounts up to 1 lb:

400 g = 1 lb
200 g = ½ lb (8 oz)
100 g = 4 oz
50 g = 2 oz
25 g = 1 oz

Liquid measures

One pint equals 0·568 litres and the metric amount has been rounded down to 0·500 litres or 500 millilitres (abbreviated to ml):

1 litre (1,000 ml) = 2 pints
½ litre (500 ml) = 1 pint
250 ml = ½ pint
125 ml = ¼ pint

Spoon measures

Metric measuring spoons have been used for the recipes in this book. These give a very accurate measurement which is especially important when measuring strong ingredients like spices. The spoons are calibrated by volume and the equivalents are as follows when the ingredient in the spoon is levelled:

15-ml spoon = 1 level tablespoon
10-ml spoon = 1 level dessertspoon
5-ml spoon = 1 level teaspoon
2·5-ml spoon = ½ level teaspoon

Fruits and vegetables best for preserving

FRUITS

Apples Late varieties are best for dry storage. Try Bramley's Seedling, Grenadier or Lane's Prince Albert. Most apples have plenty of pectin for jam- and jelly-making.
Apricots Moorpack
Blackberries Himalaya Giant or large, juicy wild fruit.
Black currants Baldwin, Boskoop, Seabrook's Black Wellington XXX or Westwick Choice
Bilberries (blueberries, blaeberries, whortleberries) Jersey
Cherries Dark red sour cherries are best for all preserves that need cooking. Choose morello or Governor Wood. Black cherries can be frozen.
Damsons Preserve well by all methods. Choose Merryweather or Shropshire Prune.
Gooseberries Green, hard fruit is best for bottling: Choose Careless and Lancer. Golden Drop, Leveller and Whinham's Industry are good for jam, jelly and freezing.
Loganberries Thornless, or Thomlus LY654.
Oranges Seville oranges make the tangiest marmalade.
Peaches Hale's Early is firm, with acidity and a freestone variety.
Pears For bottling: among dessert varieties, William's Bon Chrétien (sometimes called Bartlett), Bristol Cross, Conference or Doyenne du Comice.

For dry storage: Glou Morceau, Josephine de Malines or Winter Nelis.

For pickling and chutney: small, hard, immature dessert or cooking pears.

Pears are not recommended for freezing.
Plums For bottling: Pershore (Yellow Egg and Purple) Victoria, Warwickshire Drooper or Bryanston Gage (green) Cherry plum.

For jam: most varieties.

For freezing: Jefferson's Gage, Victoria.
Quince For jam and jelly: the common quince. The fruit of ornamental trees has little flavour but can be used for jelly.
Raspberries Glen Cova, Lloyd George, Malling Admiral, Malling Jewel, Malling Promise, Norfolk Giant or September.
Rhubarb For bottling: use young tender stems of Champagne or Linnaens.

For jams and chutney: use mature summer rhubarb.
Strawberries For jam: small strawberries are best. Cambridge Rival is a dark-coloured fruit.

For freezing: Cambridge Favourite, Cambridge Vigour, Redgauntlet, Cambridge Prizewinner, Royal Sovereign, Carnival or French Alpine.
Tomatoes Any variety but red, sun-ripened fruit has the best flavour.

For freezing: Gardener's Delight—these small fruits can be halved before freezing.

VEGETABLES

Beans, broad Aquadulce Claudia, Masterpiece Green Longpod or Promotion.
Beans, French Tendergreen, The Prince, Cordon or Masterpiece.
Beans, runner Cookham Dene, Prizewinner, Kelvedon Wonder or Fry.
Beetroot Boltardy or Little Ball.
Broccoli For freezing: Calabrese.
Brussels sprouts Peer Gynt, Citadel or Irish Elegance.
Carrots Amsterdam Forcing or Chantenay Red Cored.
Cauliflower Autumn Giant Majestic, Classic Snowball or Dominant.
Courgettes Early Gem F1 or Yellow Zucchini.
Cucumber For pickling: Burpee F1 or Gherkin.
Marrow Yellow varieties store best.
Peas For bottling and freezing Dark-skinned Perfection, Onward or Pioneer.

For drying: Harrison's Glory.
Potatoes These store well if sound.

For freezing as chips: Majestic.
Red cabbage For pickling: Large Blood Red or Red Acre.
Sweet corn For bottling or freezing: Kelvedon Glory or Canada Cross.

Preservation calendar for fruits and vegetables

January
Make marmalade, lemon and orange curds, citrus cordials, candied peel, crystallized chestnuts (marrons glacés), citrus-fruit cheeses

Freeze: Seville oranges, grapefruit segments, orange juice, lemon slices

February
Make marmalade, lemon and orange curds, citrus cordials, candied peel, citrus-fruit cheeses

Freeze: Seville oranges, grapefruit segments, orange juice, lemon slices

March
Make dried apricot jam if stocks are low
Make sharp apple chutney if apples are not storing well

Freeze: purple-sprouting broccoli
Use stored vegetables in soups if not keeping

April
Make crystallized fruits, using canned fruit, for Easter
Pickle eggs and preserve eggs in waterglass

Freeze: Lemon, orange and apricot curds; purple-sprouting broccoli

May
Bottle surplus rhubarb
Make rhubarb chutney and sauce
Bottle spinach purée and asparagus if there is a surplus

Freeze: puddings and pies with rhubarb; asparagus

June
Bottle new potatoes and gooseberries
Pickle baby beetroot
Make raspberry and strawberry jams, red-currant jelly, soft-fruit syrups
Leave peas to dry on plants
Dry herbs

Freeze: new potatoes, raspberries, strawberries, peas, broad beans, apricots

July
Make raspberry, strawberry, loganberry, blackcurrant jams and jellies; fruit liqueurs
Add raspberries, strawberries, loganberries to Rumtopf
Make soft-fruit syrups
Bottle soft fruits, carrots, peas and broad beans
Salt French and runner beans
Make piccalilli and mixed vegetable pickles, pickled red cabbage, cucumber pickle
Dry herbs

Freeze: soft fruits, apricots, beans, broccoli, carrots, courgettes

August
Make blackberry jam, jelly, butter, cheese (mix with windfall apples); damson and plum jams, jellies, butters, cheeses; mint jelly
Add damson, cherries, peaches to Rumtopf
Bottle peaches, morello cherries, plums, damsons, blackberries, apple purée (from windfalls), tomato purée, ratatouille, sweet corn
Make brandied peaches; peach brandy; sloe gin; rose hip syrup; tomato juice; red tomato chutney, sauce, ketchup; plum chutney and relish; concentrated mint sauce

Freeze: Soft fruits in pies, mousses, sorbets; morello cherries, tomatoes, courgettes, ratatouille, sweet corn

September
Make apple jam, jelly, butter, cheese (if desired, mix with soft fruits, especially blackberries); marrow jam
Add pineapple to Rumtopf
Make blackberry liqueurs; chutneys and sauces from apples, marrow, tomatoes, blackberries
Pickle plums, damsons, onions, shallots
Bottle blackberries, apple purée, ratatouille, tomatoes, mushrooms, beans
Salt beans
Dry-store cobnuts, carrots, onions, shallots, marrows

Freeze: apples as slices, as purée, in pies and desserts; blackberries, damsons, tomatoes, peppers

October
Make apple jam, jelly, butter, cheese; quince jelly and cheese; apple juice
Pickle walnuts, beetroot, onions, mixed vegetable pickles, green tomatoes
Make chutneys, sauces and ketchups from green and red tomatoes, marrow, mushrooms
Bottle apples, peppers, pears, mixed root vegetables
Dry apples
Dry-store pumpkins, marrows, onions, carrots, apples, pears

Freeze: apples as slices, as purée, in pies and desserts; tomatoes, peppers, mixed root vegetables

November
Dry-store cabbages, turnips, beetroot, celeriac, parsnips (in cold areas), chestnuts, cobnuts, walnuts
Pickle walnuts
Make glacé fruits from canned or bottled fruit for Christmas
Make marrons glacés
Candy peel for Christmas cakes and mincemeat
Make mincemeat
Make fruit liqueurs and syrups, using citrus fruits

Freeze: syrups; meat dishes, to save time for Christmas preparations

December
Candy fruit from bottled, canned and frozen fruit for presents
Make brandied jams and conserves from preserved fruit for presents
Strain and bottle Rumtopf and fruit liqueurs
Make citrus-fruit syrups, curds, cheeses, pastes
Check condition of any food stored

Freeze: Brussels sprouts in cold areas; stock from poultry carcass; surplus cooked poultry or other surplus Christmas meats

Suppliers

Look for preserving equipment in good hardware departments of large stores and shops specializing in kitchen equipment. Freezer packaging is available from freezer centres and stationers. Attractive labels can be bought in Woolworth's, Boots and W. H. Smith's.

JAM-, JELLY- AND CHUTNEY-MAKING EQUIPMENT

Attractive jam jars; jelly bags; preserving pans; funnels; and thermometers

Elizabeth David Ltd,
46 Bourne Street,
London SW1
01-730 3123

Divertimenti,
68–72 Marylebone Lane,
London W1
01-935 0689

William Page's,
91 Shaftesbury Avenue,
London W1
01-437 7021

Also Debenham's stores.

CONTAINERS AND COVERS

Plastic snap-on lids for covering jams, pickles or chutneys; parchment with waxed circles; and jelly bags

George Fowler Lee and Co. Ltd,
82 London Street,
Reading,
Berkshire,
0734 52639
(Mail order only – forty-eight-hour service.)

Plastic jars and lids

Berol Ltd,
Old Meadow Road,
Kings Lynn,
Norfolk
0553 61221

Glass 1-lb jam jars with twist-on lids

Viscose Development Co. Ltd,
185 London Road,
Croydon,
Surrey
01-686 3241

BOTTLING EQUIPMENT

Kilner jars and replacement lids

Hardware shops

Kilner: U.G. Tableware Ltd,
New Precinct, Sunbury Cross,
Sunbury-on-Thames,
Middlesex
76 87561

Other bottling jars

Elizabeth David Ltd,
46 Bourne Street,
London SW1
01-730 3123

Divertimenti,
68–72 Marylebone Lane,
London W1
01-935 0689

George Fowler Lee and Co. Ltd,
82 London Street,
Reading,
Berkshire
0734 52639

Preserving skin and other covering for jam jars and coffee jars

George Fowler Lee and Co. Ltd,
82 London Street,
Reading,
Berkshire
0734 52639

Porosan (D.I.Y.) Ltd,
P.O. Box No. 4,
57 Oving Road,
Chichester,
Sussex
0243 83781

FREEZER PACKAGING

Alcan Polyfoil Ltd,
P.O. Box 3,
Chesham,
Buckinghamshire
02405 6061

Frigicold,
Lonsdale Uniflair Ltd,
166 Dukes Road,
Western Avenue,
London W3
01-993 1271

Garfrost,
William Garfield Ltd,
Florence Street,
Birmingham
B1 1NX
021 622 2303

Lakeland Plastics,
99 Alexandra Road,
Windermere,
Cumbria
09662 2255

Porosan (D.I.Y.) Ltd,
P.O. Box No. 4,
57 Oving Road,
Chichester,
Sussex
0243 83781

Swantex Ltd,
Icefresh,
Swan Paper Co. Ltd,
Swan Mills,
Swanley,
Kent
82 65566

Thorpac Ltd,
Cirencester,

Book list and films

Gloucestershire
0285 2728

Tupperware Ltd,
P.O. Box 80,
High Wycombe,
Buckinghamshire
0494 23728

ELECTRICAL EQUIPMENT
Divertimenti,
68–72 Marylebone Lane,
London W1
01-935 0689

Kenwood Chef, with attachments for slicing, shredding and mincing.

Magimix, for chopping, grating and slicing.

SMOKING EQUIPMENT
Elizabeth David Ltd,
46 Bourne Street,
London SW1
01-730 3123

Habitat Designs Ltd,
Hithercroft Road,
Wallingford,
Oxfordshire
0491 35000

STONEWARE CROCKS
Habitat Designs Ltd,
Hithercroft Road,
Wallingford,
Oxfordshire
0491 35000

Pearsons and Co. Ltd,
The Potteries,
Whittington Moor,
Chesterfield,
Derbyshire
0246 71111

There are not many books available on the traditional methods of preservation, though there is usually a worthwhile chapter in each cookery encyclopedia. Most books on herbs and spices contain some chutney and pickle recipes. By far the largest number of books written on preservation are on freezing. Some are merely recipe books, but the ones listed below concentrate on the methods of freezing food at home with just a few examples of the special ways recipes are treated for freezing.

Complete Book of Preserving, Marye Cameron-Smith, Marshall Cavendish, 1976. £6·95.

FOR JAMS, JELLIES, PICKLES AND CHUTNEYS
Home Preservation of Fruit and Vegetables.
Her Majesty's Stationery Office, 1976, £1·00.
A very comprehensive, reasonably priced book with the scientific background to traditional home food-preservation, and a few basic recipes.

Food Storage in the Home,
R. C. Hutchinson, Edward Arnold, 1966.
A useful book of recipes and preservation information.

Let's Preserve It, Beryl Wood, Mayflower Books, 1973, 60p.
An alphabetical list of preserving recipes including many unusual ones.

Country Fare, Sheila Howarth, Pelham Books, 1976, £4·50.
Includes chapters on preservation as well as other country crafts such as bread-making.

The Penguin Book of Jams, Pickles and Chutneys,
David and Rose Mabey, Penguin Books, 1976, 75p.
An inexpensive and practical book containing many unusual recipes.

Measure for Measure
The British Diabetic Association, 3–6 Alfred Place, London WC1, 1973, £1·50.
A book of medically approved recipes for diabetics, including background information and a range of preserving recipes.

Preserves – How to Make and Use Them, Beryl Gould Marks, Faber, 1972, £2·95.

SALTING, CURING AND SMOKING
Spices, Salt and Aromatics in the English Kitchen,
Elizabeth David, Penguin Books, 1976, 90p.
A fascinating paperback of recipes for well-flavoured food including sauces and ketchups and smoked and cured food. Many of the recipes are from old cookery books.

The Home Book of Smoke Curing, Jack Sleight and Raymond Hull, David and Charles, 1973, £3·50.
A comprehensive guide to smoking, including building smoking ovens, recipes for curing meat and fish, and smoking methods.

FREEZING
Food Freezing at Home,
Gwen Conacher, The Electricity Council, 35p.

Good, sound information. An inexpensive handbook that has out-sold all other freezer books.

Freezer Facts, Margaret Leach, Forbes Publications, 1975, £4·85.
Large, detailed comprehensive volume on freezing methods.

A Guide to Home Food Freezing, Her Majesty's Stationery Office.
The companion book to *Home Preservation of Fruit and Vegetables.* It includes all the latest research on home food-freezing carried out by the Home Food Storage and Preservation Section at Long Ashton Research Station, University of Bristol.

Good Housekeeping Home Freezer Cook Book, The Good Housekeeping Institute, Ebury Press, 1972, £3·25.

The National Federation of Women's Institutes, 39 Eccleston Street, London SW1, publish a range of inexpensive booklets on homecraft subjects. The following titles give useful information on preservation:

Fruit Syrups, Cordials and Vinegar, Margaret Leach, 1976, 20p. Non-alcoholic recipes for summer drinks.

Preservation, Joan Ellis, 1975, 35p.
Informative booklet on the traditional methods of preservation.

Unusual Preserves, 1975, 35p, and **More Unusual Preserves,** 1976, 45p, by Olive Odell.
Novelty recipes for preserves using unusual fruits and vegetables.
There is also a great deal of free or very inexpensive material on food preservation which can be obtained from the following firms:

Certo-General Foods, 48 St Martin's Lane, London WC2, provide free jam, jelly and marmalade recipes.

Heinz Service for Schools, Hayes Park, Hayes, Middlesex, produces **Food Preservation,** three wall charts illustrating food preservation from early times to the present day, at 80p.

The New Zealand Lamb Information Bureau, 80–82 Cromer Street, London WC1, offers the **New Zealand lamb guide to home freezing,** and

The Meat and Livestock Commission, 5 St John's Square, London EC1, **British meat and the home freezer,** at 30p.

Films, too, on the same subjects can be acquired cheaply or on free loan.

British Sugar Bureau Film Library, 16 Paxton Place, London SE27, **Jams and preserves.**

Random Film Library, The Burroughs, Hendon, London NW4, **Endless Harvest.**

Lakeland Plastics, Alexandre Plastics, Windermere, **Freezeasy,** a 20-minute, 16-mm film on packaging for the freezer; £1·00 post and package to accompany applications.

Courses

The Home Food Storage and Preservation Section of the University of Bristol at Long Ashton Research Station hold two-day courses on traditional preservation and on freezing.

Glossary

Blanch: to scald vegetables in boiling water to kill enzymes that cause off flavours and loss of Vitamin C. Also to stop discoloration of fruits. A necessary process before freezing vegetables.
Blend: to mix ingredients together.
Brine: a salt solution used for drawing out the water from vegetables for pickling. Also used to prevent apples and pears from discolouring.
Campden tablets: when crushed and dissolved these provide a sulphite solution for preserving acid fruits. Also useful for preserving fruit syrups, cordials and wines. Available in packets from shops specializing in home-made wine supplies.

Citrus fruits: thick-skinned, juicy fruits such as oranges, lemons, grapefruit, tangerines and satsumas.
Crock: a large stone jar.
Cure: a salt and saltpetre mixture used to flavour meat.
Drip: the liquid that is released from a frozen food after thawing. A large quantity of liquid denotes a poor freezing and thawing process resulting in dry food.
Enzymes: chemical substances present in food, which speed up chemical changes. Enzymes work more quickly in warm conditions, and are destroyed by heat.
Fast-freeze switch: a switch on a home freezer to isolate the thermostat and give continuous freezing power. It is necessary to operate this switch when freezing over $\frac{1}{2}$ kg (1 lb) of food in a home freezer.
Fermentation: the conversion of sugar to alcohol and carbon dioxide, caused by the action of yeast.
Freeze: to subject food to below-zero temperatures. The water in the food forms ice crystals. If the freezing is slow, the crystals are large and the cells are ruptured.
Freezer burn: severe dehydration of the surface tissue of frozen food, avoided by correct packaging and sealing.
Grade: to sort into even sizes.
Hull: to remove stalks and leaves from berry fruit.
Kernel: the softer inside of apricot, peach and plum stones, seen when the hard outer shell has been cracked and removed.
Liqueur: a flavoured, sweetened spirit. Usually served at the end of a meal.
Mature: development of flavours, caused by storage.
Mellow: the balanced flavour achieved after storage. The flavours of the ingredients, especially the spices, blend and become more pleasant than when the preserve was first made.
Micro-organisms: minute forms of life which get into food or onto its surface. There are three classes: moulds, yeasts and bacteria.
Pare: to cut off a rind very thinly. Usually applied to citrus fruits because the white pith immediately under the skin is bitter.
Pectin: the ingredient in fruits that holds the juice in the fruit. When boiled with the right proportion of acid and sugar it forms a gel which sets jams and jellies.
Poach: to cook food in a small quantity of water at a temperature just below boiling point. It is a very gentle method of cooking that avoids the food breaking up or toughening.
Processing: cooking for a set time to sterilize bottles of fruit, vegetables, sauces, syrups and ketchups.
Pulp: fruit or vegetables cooked until almost smooth with no free liquid.
Purée: a smooth mixture from which the fibrous elements have been discarded by sieving.
Reduce: to cook a liquid by simmering or boiling until part of the liquid has evaporated.
Seal: to close with an airtight cover.
Set: to form a gel.
Shred: to cut in very thin pieces.
Sieve: to take out the fibrous part of the food by pressing it through a fine mesh.
Simmer: to cook in a liquid between 85° to 91°C (185° to 195°F). Bubbles form slowly, but collapse below the surface.
Skim: to remove any scum or particles from the surface of a liquid. A large draining spoon is used.
Skin: to remove the outer tough layer of a fruit or vegetable.
Soft fruits: the berry and currants; those with small pips and seeds, such as raspberries, strawberries, blackberries and black and red currants.
Sterilize: to kill the bacteria and enzymes that cause spoilage. This can be done with heat, or chemicals such as sulphur dioxide (see *Campden tablets*).
Stone fruits: plums, damsons, apricots, etc., with stones.
Strain: to separate the liquid from the solid parts.
Sugar syrup: made by dissolving sugar in water. Different strengths of syrup are necessary for different varieties of fruit and different purposes.
Sugar thermometer: a special thermometer which registers up to 420°F and is calibrated and marked for the different levels of sugar boiling.
Thaw: to return to an ambient temperature. Slow thawing, as in a refrigerator, causes the cells to re-absorb the moisture, and the food is then moist and succulent.
Thread: a stage in the sugar-boiling process. As sugar boils it takes on different properties until it becomes caramel. At the thread stage it will register 230°F on a sugar thermometer and will form a thread when placed on the backs of two teaspoons which are then pulled apart.
Vacuum: an airless atmosphere where bacteria and moulds that cause spoilage cannot survive.
Vitamin: a nutrient found in food. The important vitamin in fresh fruits and vegetables is Vitamin C. It is very unstable and is destroyed by soaking, by heat and by oxygen.
Zest: the thin outer skin of citrus fruits.

Index

Numbers in italics indicate captions

Credits

Author and publishers wish to thank the staff at Long Ashton Research Station, University of Bristol, for their advice on diagrams and text matter.

Artists
Vanessa Luff
Ilric Shetland
QED

Photographs
Alcan Ltd., 2
Barry Bullough/Peter Knab Associated Photographers, 12, 21, 45, 49, 56, 77
Mary Evans Picture Library, 3 (centre)
Family Circle, 3 (top), 65, 69
Paul Forrester, 29
Mansell Collection, 8
Marshall Cavendish, 32, 40
Radio Times Hulton Picture Library, 7
Ann Ronan Picture Library, 6
Syndication International, 5, 17

Pressure cooker for page 49 kindly lent by Divertimenti; glassware for page 29 by Heal's.

Cover
Design: Design Machine
Photograph: Paul Forrester